Analog IC Design

An Intuitive Approach

8th Edition

By

Gabriel Alfonso Rincón-Mora

School of Electrical and Computer Engineering
Georgia Institute of Technology

Rincon-Mora.gatech.edu

V. 2

Discovering the universe through the art of design

Contents

Preface

I. Intended Audience

Integrated-circuit (IC) design, system, and product engineers, managers, and undergraduate and graduate engineering students engaged or interested in expanding their knowledge on how to analyze, design, evaluate, specify, develop, and test analog ICs.

II. Description and Objectives

This slide book presents, explains, and shows how to understand, develop, and use semiconductor devices to model, analyze, and design transistor-level analog integrated circuits (ICs) with and without feedback using bipolar and CMOS technologies. The underlying aim is to cultivate and develop *insight* and *intuition* for how semiconductor devices work individually and collectively in microelectronic circuits. For this, the presentation seeks to furnish an intuitive view of ICs that transcends mathematical and algebraic formulations to empower engineers with the tools necessary to design ICs that perform practical and complex analog functions.

G.A.R.M.

Atlanta, Georgia

List of Abbreviations & Variables

BJT: Bipolar-junction transistor
CB: Common-base transistor/stage
CC: Common-collector transistor/stage
CD: Common-drain transistor/stage
CE: Common-emitter transistor/stage
CG: Common-gate transistor/stage
CMRR: Common-mode rejection ratio
CS: Common-source transistor/stage
CTAT: Complementary to absolute temperature
ESR: Equivalent series resistance
FB: Feedback
FET: Field-effect transistor
FW: Forward
GBW: Gain–bandwidth product
GM: Gain margin
HD: Harmonic distortion
ICMR: Input common-mode range
JFET: Junction FET
KCL: Kirchhoff's current law
KVL: Kirchhoff's voltage law/loop
LDR: Load regulation
LNR: Line regulation
MOS: Metal–oxide–semiconductor
NFET: N-channel MOSFET
NMOS: N-channel MOSFET
OC: Open-circuit model
PFET: P-channel MOSFET
PM: Phase margin
PMOS: P-channel MOSFET
PK: Peak
PSR: Power-supply rejection
PSRR: PSR ratio/power-supply ripple rejection
PTAT: Proportional to absolute temperature
RHP: Right-half plane
RMS: Root–mean–squared
RSS: Root–sum–squared
SC: Shunt/short-circuit model
SNR: Signal-to-noise ratio
SR: Slew rate
TC: Fractional temperature coefficient
THD: Total harmonic distortion

A_β: Feedback-path gain
A_{CL}: Closed-loop gain
A_D: Differential gain

A_C: Common-mode gain
A_{DD}: Positive power-supply gain
A_E: Emitter area/Error-amplifier gain
A_F: Forward-path gain
A_{FW}: Forward gain
A_G: Transconductance gain
A_I: Current gain (translation)/Amplitude of ith harmonic
A_{IN}: Input-supply gain
A_{IO}: Input-to-output gain
A_J: PN junction area
A_{LG}: Loop gain
$\angle A_{LG}$: A_{LG}'s phase shift
A_{LNR}: Line-regulation gain
A_N: Norton gain
A_P: Pass-switch gain
A_{PRE}: Pre-amplifier gain
A_{SS}: Negative power-supply gain
A_T: Thévenin gain
A_V: Voltage gain
A_{X0}: A_X at low frequency
$A_{X(0dB)}$: A_X at f_{0dB}
$A_{X(180°)}$: A_X at $f_{180°}$
$A_{X(HF)}$: A_X at high frequency
A_Z: Transimpedance gain

β_0: BJT's base–collector current gain
β_{FB}: Feedback translation

C_{BC}: Base–collector capacitance
C_{BD}: Body–drain capacitance
C_{BE}: Base–emitter capacitance
C_{BS}: Body–source capacitance
C_{CH}: Channel capacitance
C_{CL}: Equivalent closed-loop capacitance
C_{DEP}: Depletion capacitance
C_{DIF}: Diffusion capacitance
C_G: Gate capacitance
C_{GD}: Gate–drain capacitance
C_{GS}: Gate–source capacitance
C_{IN}: Input capacitance
C_J: PN junction capacitance
C_{J0}: Zero-bias C_J
C_{J0}'': C_{J0} per unit area
C_{LD}: Load capacitance
C_O: Output capacitance
C_{OL}: Overlap capacitance
C_{OX}: Oxide capacitance
C_{OX}'': C_{OX} per unit area

C_{SUB}: Substrate capacitance
C_T: Tail-trainsistor capacitance
C_μ: Base–collector capacitance
C_π: Base–emitter capacitance

d_W: Depletion width

E_V: Voltage error
ε_0: Permittivity in vacuum
ε_{OX}: Oxide permittivity
ε_{Si}: Relative permittivity of silicon

f_{0dB}: Unity-gain frequency
$f_{180°}$: Inversion frequency
f_{BW}: Bandwidth (–3-dB) frequency
f_O: Operating frequency
f_T: Transitional frequency
f_X: Shunting RC pole at node v_X
f_X': f_X when finding p_1
f_X'': f_X when finding p_2

G_B: Buffer transconductance
G_C: Common-mode transconductance
G_D: Differential transconductance
g_{ds}: Small-signal drain–source conductance
G_E: Error-amplifier transconductance
G_{FB}: Feedback transconductance
G_{FW}: Forward transconductance
G_{LG}: Loop-gain transconductance
g_m: Small-signal mutual transconductance
G_M: Mutual transconductance
g_m': Effective g_m
g_{mb}: MOS body's g_m
g_o: Small-signal output conductance
G_O: Output transconductance
G_P: Pass-switch transconductance
g_π: Small-signal base–emitter conductance

HD_i: Harmonic distortion of ith harmonic

i_A: Amplifier current
i_B: Base/buffer current
I_B: Bias current/static component of i_B
i_C: Collector/CTAT current
i_D: Diode/drain current
i_E: Emitter current
i_{DD}: v_{DD} current
i_F: Forward current

i_{FB}: Feedback current
i_{FW}: Forward current
i_g: g_m current
i_G: Gate current
i_{LD}: Load current
i_n^*: Non-degenerated noise current
i_N: Norton current
I_{OS}: Offset current
$I_{OS(S)}$: Systemic I_{OS}
I_{OS}^*: Random I_{OS}
i_P: PTAT/pass-switch current
i_Q: Quiescent current
i_R: Reverse current
i_S: Source current
I_S: Reverse-saturation current/static component of i_S
i_{SS}: v_{SS} current
i_{ST}: Starter current
i_T: Tail-transistor current
i_π: r_π current

$K_{N/P}'$: N/P-channel MOS transconductance parameter
K_B: Boltzmann's constant
K_D: Initial drive fraction
K_O: Overdrive factor

L_{CH}: Channel length
L_{MIN}: Minimum oxide length
L_{OL}: Overlap length
$\lambda_{N/P}$: N/P channel-length modulation parameter

$\mu_{N/P}$: Electron/hole mobility

n_I: Ideality/non-ideality factor
η_X: Power-conversion efficiency for device "X"

p_1: Lowest-frequency pole
p_2: 2nd lowest-frequency pole
P_{GND}: Ground power
P_A: Amplifier power
p_B: Buffer/base pole
p_{BW}: Bandwidth pole
p_C: Capacitor pole
p_{CX}: Reversal capacitor pole
P_{DD}: v_{DD} power
p_{DX}: Reversal degeneration pole
P_{SS}: v_{SS} power
p_E: Error-amplifier pole
p_F: Folding pole

p_{IN}: Input pole
p_{LD}: Load pole
P_{LOSS}: Power loss
p_M: Mirror pole
p_O: Output pole
P_O: Output power
P_R: Ohmic power
p_X: Shunting pole at node v_X

q_{BC}: Base–collector charge
q_D: Diode charge
q_{DEP}: Depletion charge
q_{DIF}: Diffusion charge
q_E: Electronic charge
q_{FR}: Forward-recovery charge
q_{RR}: Reverse-recovery charge

R_B: Equivalent base resistance
R_C: Equivalent collector resistance/capacitor's resistance
R_{CH}: Channel resistance
R_{CL}: Closed-loop resistance
r_d: Small-signal diode resistance
R_D: Equivalent drain resistance
R_{DO}: Dropout resistance
r_{ds}: Small-signal drain–source resistance
R_E: Equivalent emitter resistance
R_{FB}: Feedback resistance
r_{gm}: Equivalent g_m resistance
R_{IC}: Common-mode input resistance
R_{ID}: Differential input resistance
R_{IN}: Input resistance
R_{LD}: Load resistance
R_N: Norton resistance
R_{OL}: Open-loop resistance
r_o: Small-signal output resistance
R_O: Output resistance
R_{OC}: Common-mode output resistance
R_{OD}: Differential output resistance
R_S: Equivalent source resistance
R_{SH}: Shunt resistance
R_T: Tail-transistor resistance/Thévenin resistance
r_π: Small-signal base–emitter resistance

s_E: Error signal
s_{FB}: Feedback signal
s_I: Input signal
s_{IN}: Input signal
s_O: Output signal

S_X: Width-to-length (aspect) ratio of MOSFET M_X
$\gamma_{N/P}$: N/P-channel body-effect parameter

T_J: Junction temperature
t_{OX}: Oxide thickness
t_P: Propagation delay
t_R: Response time
T_{ROOM}: Room temperature
t_{SR}: SR delay
τ_{BW}: Bandwidth time constant
τ_F: Forward transit time
τ_L: Latch's time constant
τ_T: Transit time
τ_X: Time constant for stage X

V_A: Early (base-width-modulation) voltage
v_B: Base/buffer voltage
v_{BAT}: Battery voltage
v_{BC}: Base–collector voltage
V_{BD}: Breakdown voltage
v_{BE}: Base–emitter voltage
V_{BG}: Bandgap voltage
V_{BI}: Built-in voltage
v_{BS}: Body–source voltage
v_{CC}: Positive power-supply voltage
v_{CE}: Collector–emitter voltage
v_D: Diode/drain voltage
v_{DD}: Positive power-supply voltage
v_{DO}: Dropout voltage
v_{DS}: Drain–source voltage
v_E: Error voltage
v_{EE}: Negative power-supply voltage
v_{FB}: Feedback voltage
v_{HYS}: Hysteresis voltage
v_I: Input voltage
v_{IN}: Input voltage
v_{IC}: Common-mode input voltage
v_{ID}: Differential input voltage
V_{IH}: Input-high voltage
V_{IL}: Input-low voltage
v_G: Gate voltage
v_{GS}: Gate–source voltage
v_{GST}: Gate drive
v_{LD}: Load voltage
v_N: Negative input
v_O: Output voltage
V_{OH}: Output-high voltage
V_{OL}: Output-low voltage

v_{OUT}: Output voltage
V_{OS}: Offset voltage
$V_{OS(S)}$: Systematic V_{OS}
$V_{OS}{}^{*}$: Random V_{OS}
v_P: Positive input
v_R: Reference/resistor voltage
v_S: Source voltage
v_{SS}: Negative power-supply voltage
v_{SUB}: Substrate voltage
V_t: Thermal voltage
v_T: Threshold/Thévenin voltage
v_{TH}: Threshold voltage
V_{T0}: Zero-bias v_T
$v_x{}^{*}$: Noise voltage at node v_X
ψ_B: Surface barrier potential

W_{CH}: Channel width
W_{MIN}: Minimum W_{CH}

z_C: Capacitor zero
z_{CX}: Reversal capacitor zero
z_D: Degeneration zero
Z_{FB}: Feedback impedance
Z_G: Impedance to ground
z_{GD}: Gate–drain zero
z_{GS}: Gate–source zero
Z_{LD}: Load impedance
z_M: Mirror zero
z_{MX}: Reversal mirror zero
Z_O': Output impedance for A_{LG} response
Z_O'': Output impedance for A_{IN} response
Z_{SER}: Series impedance
Z_{SH}: Shunt impedance

Chapter 1. Analog Electronics

1.1. Two-Port Models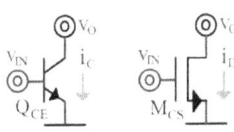

1.2. Electronic Devices

1.3. Transistor Circuits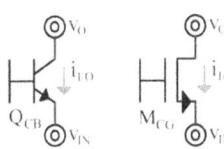

1.4. Frequency Response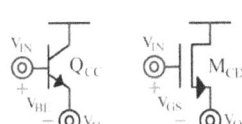

1.1. Two-Port Models: Extraction

Purpose: Predict loaded response with simple model

Principle: Orthogonal (mutually independent) components

Extraction: Nullify other components

Thévenin: 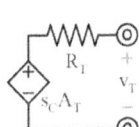   Derive A_T w/o R_T $\rightarrow$ $v_R \equiv 0$ $\therefore$ $i_T \equiv 0$ $\rightarrow$ Remove load

Derive R_T w/o A_T $\therefore$ $s_C \equiv 0$

Norton: 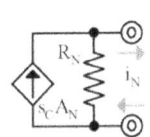   Derive A_N w/o R_N $\rightarrow$ $i_R \equiv 0$ $\therefore$ $v_N \equiv 0$ $\rightarrow$ Short output

Derive R_N w/o A_N $\therefore$ $s_C \equiv 0$

Bidirectional Example: Reverse Hybrid   Forward Example: No Feedback

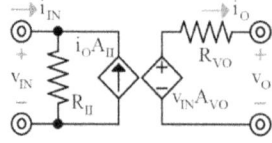

i-Sourced v_O →

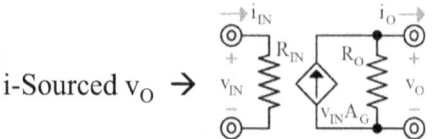

Feedback Model: Derive R_{II} when $i_O \equiv 0$ ∴ Remove load

Derive A_{II} when $i_{RI} \equiv 0$ ∴ $v_{IN} \equiv 0$ → Short input

Derive A_{VO} when $v_{RO} \equiv 0$ ∴ $i_O \equiv 0$ → Remove load

Derive R_{VO} when $v_{IN} \equiv 0$ ∴ Short input

Note: $A_{I/O}$ = Feedback/forward translations $R_{I/O}$ test conditions disable $A_{I/O}$

→ $A_{I/O}$ & $R_{I/O}$ = Open-loop parameters that model closed-loop behavior

Forward Model: Derive R_{IN} as is   → No redundancies to consider

Derive A_G when $i_{RO} \equiv 0$ ∴ $v_O \equiv 0$ → Short output

Derive R_O when $v_{IN} \equiv 0$ ∴ Short input

1.2. Electronic Devices: A. Diodes: i. Operation

Regions

Zero Bias: $v_D = 0$

Forward Bias: $v_D > 0$

Reverse: $v_D < 0$

Breakdown: $v_D \approx -V_{BD}$

Parameters:

Built-in Barrier Voltage ≡ $V_{BI} \approx 600$ mV

Reverse-Saturation Current ≡ I_S

Thermal Voltage ≡ $V_t \approx 26$ mV at 27°C

Non-Ideality Factor ≡ $\eta_I = 1–2$

Zener   Schottky

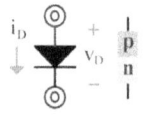
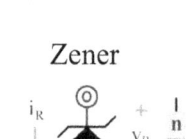
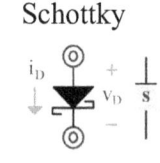

$$i_D = I_S \left[\exp\left(\frac{v_D}{\eta_I V_t}\right) - 1 \right] \propto A_J$$

$$\approx I_S \exp\left(\frac{v_D}{\eta_I V_t}\right) \text{ When } v_D > 3V_t$$

Junction Area ≡ A_J

ii. Dynamic Response

Small Variations: Zero-v_D Junction Capacitance $\equiv C_{J0}$

$$C_J = \frac{\Delta q_D}{\Delta v_D} = \frac{\Delta i_D \Delta t_R}{\Delta v_D} = C_{DEP} + C_{DIF}$$

C_{J0} per unit Area $\equiv C_{J0}"$

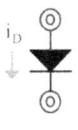

$$C_{DEP} = \frac{C_{J0}}{\sqrt{\dfrac{V_{BI} - v_D}{V_{BI}}}} = \frac{A_J C_{J0}"}{\sqrt{\dfrac{V_{BI} + v_R}{V_{BI}}}} \propto \frac{A_J}{d_W}$$

Transit Time $\equiv \tau_T$

Small-signal conductance
$\downarrow$

$$C_{DIF} = \frac{\Delta q_{DIF}}{\Delta v_D} = \frac{\Delta i_D \tau_T}{\Delta v_D} \approx \left(\frac{\partial i_D}{\partial v_D}\right)\tau_T \equiv g_d \tau_T \approx \left(\frac{i_D}{\eta_I V_t}\right)\tau_T$$

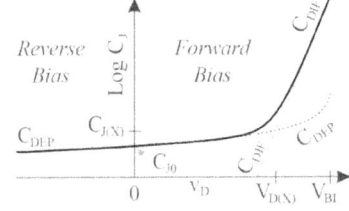

Large Variations: Forward/Reverse Recovery

Feed/pull $q_{DIF} + q_{DEP}$ $\rightarrow$ $\quad t_{F/RR} = \dfrac{q_{DIF} + q_{DEP}}{i_D} = \dfrac{i_F \tau_T + C_{DEP}\Delta v_D}{i_{FR}} \approx \dfrac{i_F \tau_T + C_{J0}(v_F + v_R)}{i_{FR}}$

Schottky: $V_{BI} \approx 300\text{--}500$ mV (usually < PN diode) No q_{DIF} $\rightarrow$ Short $t_{F/RR}$

B. BJTs: i. Large Signal

Regions

Active: $v_{BE} > 0$ $v_{BC} < 0$

Saturation: $v_{BC} > 0$ $\rightarrow$ $v_{CE} < v_{BE}$

Light: $v_{CE} > v_{CE(MIN)}$

Deep: $v_{CE} \leq v_{CE(MIN)} \approx \underbrace{v_{BE} - v_{BC(5\%iC)}}_{} + i_C R_C \approx 200\text{--}300$ mV

Reverse: $v_{BC} > 0$ $v_{BE} < 0$ $\approx 3\eta_I V_t$

NPN: $i_C \approx I_S \left[\exp\left(\dfrac{v_{BE}}{\eta_I V_t}\right) - 1\right]\left(1 + \dfrac{v_{CE}}{V_A}\right) \propto A_{JBE}$ $i_E = i_B + i_C$

PNP: $i_C \approx I_S \left[\exp\left(\dfrac{v_{EB}}{\eta_I V_t}\right) - 1\right]\left(1 + \dfrac{v_{EC}}{V_A}\right) \propto A_{JBE}$ $i_B = \dfrac{i_C}{\beta_0}$

Base-Width Modulation $\equiv V_A$ Base–Emitter Area $\equiv A_{JBE}$ Current Gain $\equiv \beta_0$

ii. Capacitances

Small Variations: Forward $\tau_T \equiv \tau_F$

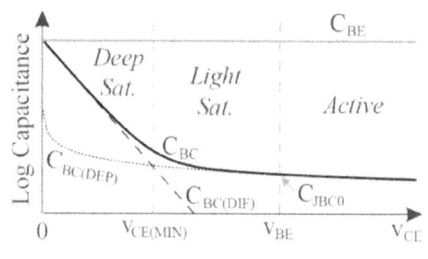

$$C_\pi\Big|_{v_{BE}\leq 0}^{v_{BE}<300-500mV} \approx C_{BE(DEP)} = \frac{A_{JBE}C_{JBE0}{''}}{\sqrt{\dfrac{V_{BI}-V_{BE}}{V_{BI}}}}$$

$$C_\pi\Big|_{v_{BE}>300-500mV} \approx C_{BE(DIF)} \approx \left(\frac{\partial i_C}{\partial v_{BE}}\right)\tau_F \approx \left(\frac{I_C}{\eta_I V_t}\right)\tau_F \equiv g_m\tau_F$$

$$C_{BE} \equiv C_\pi$$

$$C_{BC} \equiv C_\mu$$

$$C_\mu\Big|_{v_{CE}\geq v_{BE}}^{v_{CE}\geq v_{CE(MIN)}} \approx C_{BC(DEP)} = \frac{A_{JBC}C_{JBC0}{''}}{\sqrt{\dfrac{V_{BI}+v_{CB}}{V_{BI}}}} = \frac{A_{JBC}C_{JBC0}{''}}{\sqrt{\dfrac{V_{BI}+v_{CE}-v_{BE}}{V_{BI}}}}$$

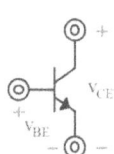

Large Variations:

   Active Recovery: Pull $q_{BC(DIF)} + q_{DEP}$ requires reverse i_B & t_{AR}

   Saturation Recovery: Feed $q_{BC(DIF)} + q_{DEP}$ requires forward i_B & t_{SR}

C. MOSFETs: i. N Channel

Regions

Accumulation $\equiv$ Cut Off: $v_{GS} < 0$

Depletion $\equiv$ Sub-v_T: $0 < v_{GS} < v_{TN}$

   Triode: $v_{DS} \leq 3V_t$ $i_D\Big|_{v_{BE}\leq 0}^{0<v_{GS}\leq v_{TN}} = \left(\dfrac{W}{L}\right)I_{SN}\exp\left(\dfrac{v_{GS}-v_{TN}}{\eta_I V_t}\right)\left[1-\dfrac{1}{\exp\left(v_{DS}/V_t\right)}\right]$

   Saturation: $v_{DS} \geq v_{DS(SAT)}{}' \approx 3V_t$

Parameters: $i_D\Big|_{v_{DS}\geq 3V_t}^{0<v_{GS}\leq v_{TN}} \approx \left(\dfrac{W}{L}\right)I_{SN}\exp\left(\dfrac{v_{GS}-v_{TN}}{\eta_I V_t}\right)$

Channel Width & Length $\equiv$ W & L

Saturation Current $\equiv I_{SN}$

Threshold Voltage $\equiv v_{TN} \approx 400-600$ mV

Non-Ideality Factor $\equiv \eta_I \approx 1-2$

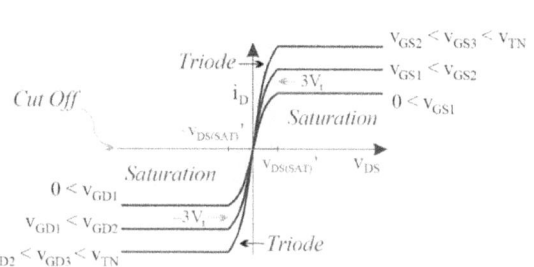

Inversion: $v_{GS} > v_{TN}$

Triode → $v_{GD} \geq v_{TN}$ → $v_{DS} \leq v_{DS(SAT)}$

Sat. → $v_{GD} < v_{TN}$ → $v_{DS} > v_{DS(SAT)}$

$$R_{CH}\Big|_{v_{DS} \ll v_{GST}}^{v_{GS} > v_{TN}} \approx \left(\frac{L}{W}\right)\left[\frac{1}{K_N'(v_{GS} - v_{TN})}\right]$$

$$v_{TN} = V_{TN0} + \gamma_N\left(\sqrt{2\psi_B - v_{BS}} - \sqrt{2\psi_B}\right)$$

Surface–Body Barrier Potential $\equiv \psi_B$

$$i_D\Big|_{v_{DS} \leq v_{GST}}^{v_{GS} > v_{TN}} = \frac{v_{DS}}{R_{CH}} = v_{DS}\left(\frac{W}{L}\right)K_N'\left(v_{GS} - v_{TN} - \frac{v_{DS}}{2}\right)$$

$$K_N' = \mu_N C_{OX}'' = \mu_N\left(\frac{\varepsilon_{OX}}{t_{OX}}\right) = \mu_N\left(\frac{3.9\varepsilon_0}{t_{OX}}\right)$$

$$i_D\Big|_{v_{DS} > v_{GST}}^{v_{GS} > v_{TN}} = \frac{v_{DS(SAT)}}{R_{CH}} \approx \left(\frac{W}{L}\right)\left(\frac{K_N'}{2}\right)(v_{GS} - v_{TN})^2(1 + \lambda_N v_{DS})$$

$$v_{DS(SAT)} = v_{GS} - v_{TN} \equiv v_{GST}$$

Parameters: Zero-v_{BS} $v_{TN} \equiv V_{TN0}$ Body-Effect Parameter $\equiv \gamma_N \approx 0.4\text{–}0.6 \sqrt{V}$

Transconductance Parameter $\equiv K_N'$ Channel-Length Modulation Parameter $\equiv \lambda_N$

Electron Mobility $\equiv \mu_N$ Oxide Capacitance/Area $\equiv C_{OX}''$

Example: Determine R_{CH} when $W_{CH} = 10$ μm, $L_{CH} = 170$ nm, $v_{GS} = 1.8$ V,

$v_{DS} = 50$ mV, $\mu_N = 72k$ mm²/V·s, $t_{OX} = 15$ nm, $v_{TN} = 400$ mV.

Solution: $C_{OX}'' = \dfrac{\varepsilon_{OX}}{t_{OX}} = \dfrac{\varepsilon_{Si}\varepsilon_0}{t_{OX}} = \dfrac{3.9(8.845\ \text{pF/m})}{15\ \text{nm}} - 2.3\ \text{mF/m}^2 = 2.3\ \text{fF/μm}^2$

$K_N' = \mu_N C_{OX}'' = (72m)(2.3m) = 170\ \text{μA/V}^2$

$v_{DS} = 50$ mV $\ll v_{GST} = v_{GS} - v_{TN} = 1.8 - 400m = 1.4$ V

∴ Deep in Triode → $R_{CH} \approx \left(\dfrac{L_{CH}}{W_{CH}}\right)\left[\dfrac{1}{K_N'(v_{GS} - v_{TN})}\right] = 71\ \Omega$

Example: Determine v_{TN} when $V_{TN0} = 400$ mV, $v_{BS} = -100$ mV, $\gamma_N = 600$ m√V,

$\psi_B = 300$ mV.

Solution: $v_{TN} = V_{TN0} + \gamma_N\left(\sqrt{2\psi_B - v_{BS}} - \sqrt{2\psi_B}\right)$

$= 400m + (600m)\left(\sqrt{2(300m) - (-100m)} - \sqrt{2(300m)}\right) = 440$ mV

ii. P Channel

Regions

$v_G > v_S$

↓

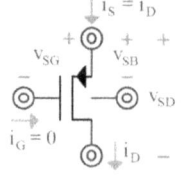 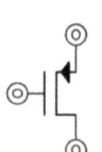 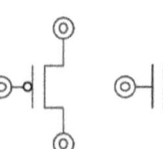

Accumulation ≡ Cut Off: $v_{SG} < 0$

Depletion ≡ Sub-v_T: Triode Saturation

Inversion: Triode Saturation $i_D\big|^{0<v_{SG}\leq|v_{TP}|} = \left(\dfrac{W}{L}\right) I_{SP} \exp\left(\dfrac{v_{SG}-|v_{TP}|}{\eta_I V_t}\right)\left[1 - \dfrac{1}{\exp\left(v_{SD}/V_t\right)}\right]$

$R_{CH}\Big|^{v_{SG}>|v_{TP}|}_{v_{SD}\ll v_{SGT}} \approx \left(\dfrac{L}{W}\right)\left[\dfrac{1}{\mu_P C_{OX}{}''\left(v_{SG}-|v_{TP}|\right)}\right]$

$i_D\Big|^{0<v_{SG}\leq|v_{TP}|}_{v_{SD}\geq 3V_t} \approx \left(\dfrac{W}{L}\right) I_{SP} \exp\left(\dfrac{v_{SG}-|v_{TP}|}{\eta_I V_t}\right)$

$i_D\Big|^{v_{SG}>|v_{TP}|}_{v_{SD}\leq v_{SGT}} = v_{SD}\left(\dfrac{W}{L}\right) K_P'\left(v_{SG}-|v_{TP}|-\dfrac{v_{SD}}{2}\right)$

$|v_{TP}| = |V_{TP0}| + \gamma_P\left(\sqrt{2\psi_B - v_{SB}} - \sqrt{2\psi_B}\right)$

$v_{SD(SAT)} = v_{SG} - |v_{TP}| \equiv v_{SGT}$

$i_D\Big|^{v_{SG}>|v_{TP}|}_{v_{SD}>v_{SGT}} \approx \left(\dfrac{W}{L}\right)\left(\dfrac{K_P'}{2}\right)\left(v_{SG}-|v_{TP}|\right)^2\left(1+\lambda_P v_{SD}\right)$

$\approx \sqrt{\dfrac{2i_D}{(W/L)K_P'\left(1+\lambda_P v_{SD}\right)}}$

Weak Inversion: $v_{GS} \approx v_T \pm 50$ mV

Conduction: Carriers diffuse & drift

Small-Signal Transconductance:

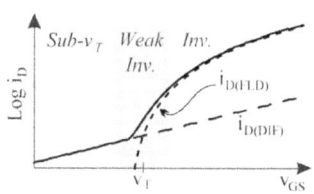

$g_m\Big|^{0<v_{GS}<v_T}_{v_{DS}\geq 3V_t} \equiv \dfrac{\partial i_D}{\partial v_{GS}} \approx \left(\dfrac{1}{\eta_I V_t}\right)\left(\dfrac{W}{L}\right) I_s \exp\left(\dfrac{v_{GS}-v_T}{\eta_I V_t}\right) \approx \dfrac{i_D}{\eta_I V_t}$ $v_{DS(SAT)} = 2\eta_I V_t$

$g_m\Big|^{v_{GS}>v_T}_{v_{DS}>v_{GST}} \equiv \dfrac{\partial i_D}{\partial v_{GS}} \approx \left(\dfrac{W}{L}\right) K'\left(v_{GS}-v_T\right)\left(1+\lambda v_{DS}\right)$

$= \sqrt{2i_D K'(W/L)\left(1+\lambda v_{DS}\right)}$

$g_{m(MAX)}\Big|^{v_{GS}>v_T}_{v_{DS}>v_{GST}} \approx \sqrt{2i_D K'(W/L)_X\left(1+\lambda v_{DS}\right)} = g_m\Big|^{0<v_{GS}<v_T}_{v_{DS}\geq 3V_t} \approx \dfrac{i_D}{\eta_I V_t}$

$\therefore \left(\dfrac{W}{L}\right)_X \approx \dfrac{i_D}{2\eta_I^2 V_t^2 K'\left(1+\lambda v_{DS}\right)}$ → $v_{DS(SAT)}\Big|^{v_{GS}>v_T}_{(W/L)_X} \approx \sqrt{\dfrac{2i_D}{(W/L)_X K'\left(1+\lambda v_{DS}\right)}} = 2\eta_I V_t$

iii. Capacitances

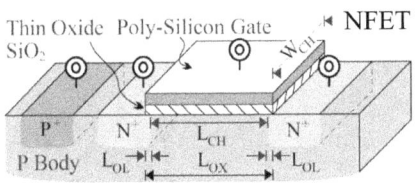

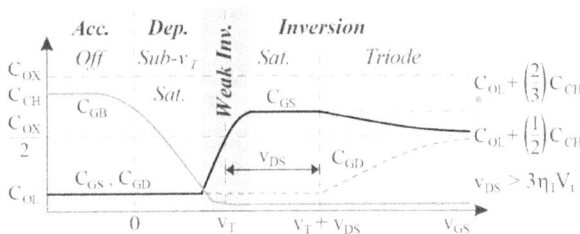

PN Junctions: C_{SB} C_{DB}

Gate Oxide: C_{GS} C_{GB} C_{GD}

Cut Off C_{OL} C_{CH} C_{OL}

Sub-v_T Channel depletes

Triode Inv. $C_{GS}:C_{GD}$ share C_{CH}

Sat. Inv. C_{GS} holds $(2/3)C_{CH}$

$$C_J \approx C_{DEP} = \frac{A_J C_{J0}''}{\sqrt{\dfrac{V_{BI} + V_{JR}}{V_{BI}}}}$$

$$L_{OX(MIN)} \equiv L_{MIN}$$

$$C_{OL} = C_{OX}''W_{CH}L_{OL} \quad \rightarrow \quad L_{OL} \approx \frac{L_{OX(MIN)}}{6}$$

$$C_{CH} = C_{OX}''W_{CH}L_{CH} \quad \rightarrow \quad L_{CH} = L_{OX} - 2L_{OL}$$

$$C_{GB} = C_{CH} \oplus C_{DEP} \leq Min\{C_{CH}, C_{DEP}\}$$

Example: Determine C_G's when W = 10 μm, L = 250 nm, L_{OL} = 40 nm,

C_{OX}'' = 2.3 fF/μm², V_{TN0} = 400 mV, v_D = 400 mV, v_G = 700 mV, $v_S = v_B = 0$ V.

Solution:

$C_{OL} = C_{OX}''W_{CH}L_{OL} = 0.92$ fF

$C_{CH} = C_{OX}''W_{CH}(L_{OX} - 2L_{OL}) = 3.9$ fF

$v_{GS} = v_G - v_S = 700$ mV $> v_T = V_{TN0} = 400$ mV $\quad \therefore \quad$ Inverted

$v_{DS} = v_D - v_S = 400$ mV $> v_{DS(SAT)} = v_{GS} - v_T = 300$ mV $\quad \therefore \quad$ Saturated

$C_{GS} \approx C_{OL} + \left(\dfrac{2}{3}\right)C_{CH} = 3.5$ fF $\qquad C_{GD} = C_{OL} = 0.92$ fF $\qquad C_{GB} \leq C_{DEP}$

D. Small-Signal Models

Diode:

$$i_D = f(v_D)$$

$$g_d \equiv \frac{1}{r_d} \equiv \frac{\partial i_D}{\partial v_D} \approx \frac{I_D}{V_t}$$

BJT:

$$i_B = f(v_{BE})$$

$$i_C = f(v_{BE}, v_{CE})$$

$$g_o = \frac{1}{r_o} \approx \frac{I_C}{V_A} \qquad r_\pi \approx \frac{\beta_0}{g_m} = \beta_0 r_{gm}$$

$$g_m \equiv \frac{1}{r_{gm}} \approx \frac{I_C}{V_t} \qquad f_T = \frac{g_m}{2\pi(C_\pi + C_\mu)}$$

MOSFET:

$$i_G = 0$$

$$i_D = f(v_{GS}, v_{BS}, v_{DS})$$

$$g_{m(SUB)} \approx \frac{I_D}{n_I V_t} \qquad g_{mb} \approx \frac{\gamma_N g_m}{2\sqrt{2\psi_B - V_{BS}}}$$

$$g_{m(INV)} \approx \sqrt{2 I_D K_{NP}'(W_{CH}/L_{CH})(1 + \lambda v_{DS})}$$

$$(0\text{-}v_{BS}) \text{ Transitional Frequency} \equiv f_T = \frac{g_m}{2\pi(C_{GS} + C_{GD})} \qquad g_{ds} = \frac{1}{r_{ds}} \approx I_D \lambda_{NP}$$

E. MOS Design

Bandwidth:

$$f_{T(SUB)} \propto \frac{g_{m(SUB)}}{C_{EQ}} \propto \frac{1}{W_{CH} L_{OX}}$$

$$f_{T(INV)} \propto \frac{g_{m(INV)}}{C_{EQ}} \propto \frac{\sqrt{W_{CH}/L_{OX}}}{W_{CH} L_{OX}} = \frac{1}{W_{CH}^{0.5} L_{OX}^{1.5}}$$

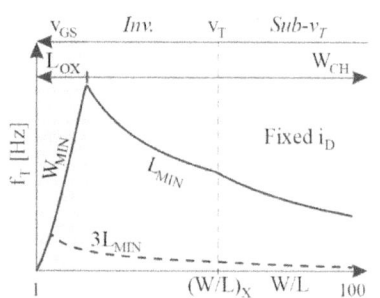

High Gain & Bandwidth: Highest I_D (that P_Q allows)

$\qquad g_m \qquad\qquad f_T$

$\qquad\qquad\qquad \to \; V_{GS} \propto I_D^X \; \therefore \; $ Raise I_D towards strong inversion

$\qquad\qquad\qquad$ Shortest L_{OX} (that matching & λ in r_{ds} in A_V allow)

Limit: $\quad$ Higher W_{CH} raises $g_{m(INV)}$ & reduces f_T $\quad\rbrack \quad W_{CH} \leq W_X = (W/L)_X L_{CH}$

$\qquad\qquad g_m$ peaks in sub-v_T $\qquad\qquad\qquad\qquad\quad\rbrack \quad$ Sub-v_T Transition Level

Square-Root $g_{m(INV)}$ < n_I-Suppressed Exponential $g_{m(SUB)}$ < Exponential $g_{m(BJT)}$

1.3. Transistor Circuits

Impedance Combinations

Parallel: $\qquad R_A \parallel R_B \le \text{Min}\ \{R_A, R_B\} \qquad\qquad C_A \parallel C_B \equiv C_A + C_B$

Series: $\qquad R_A \oplus R_B \equiv R_A + R_B \qquad\qquad C_A \oplus C_B \le \text{Min}\ \{C_A, C_B\}$

Signal Composition

Bias $\equiv$ Static components: $\qquad$ Upper-case variables $\rightarrow$ I_C, V_{DS}, etc.

Small-signal components: $\qquad$ Lower-case variables $\rightarrow$ i_d, v_{be}, etc.

Complete signals: Lower-case variables with upper-case subscripts $\rightarrow$ i_S, v_D, etc.

Terminal Roles

$i_{B/G} \approx$ Very low $\qquad\qquad\qquad \therefore \qquad$ B, G = Not useful as outputs

$i_{C/D} \approx i_{E/S} \gg i_{B/G}$ $\qquad\qquad \therefore \qquad$ C, D, E, S = Useful outputs

$i_{C/D}$ = Sensitive to $v_{BE/GS}$ $\qquad\quad \therefore \qquad$ B, E, G, S = Useful inputs

$i_{C/D} \approx$ Usually insensitive to $v_{C/D}$ $\qquad \therefore \qquad$ C, D = Not useful as inputs

A. Primitives

Input & output use 2 terminals $\rightarrow$ Unused Terminal $\equiv$ Common Terminal

Common (unused) Emitter/Source:

$\qquad$ C/D = Not useful as input

$\qquad \therefore$ C/D = Output $\quad$ B/G = Input

Common (unused) Base/Gate:

$\qquad$ C/D = Not useful as input

$\qquad \therefore$ C/D = Output $\quad$ E/S = Input

Common (unused) Collector/Drain:

$\qquad$ B/G = Not useful as output

$\qquad \therefore$ B/G = Input $\quad$ E/S = Output

Primary Roles:

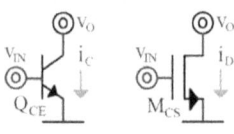

Transconductor

CE/S: Δv_{IN} → Substantial $\Delta i_{C/D}$

   ∴   Transconductor

CB/G: $i_{C/D} \approx i_{E/S}$

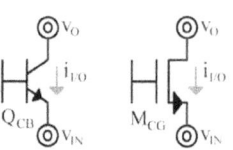

Current Buffer

   ∴   Current Buffer

CC/D: If $i_{C/D} \approx I_{C/D}$ → $v_{BE/GS} \approx V_{BE/GS} \approx$ Constant

   ∴ $\Delta v_O \approx \Delta v_{IN}$ → Voltage Follower

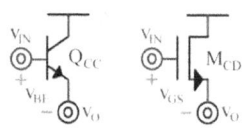

Voltage Follower

Typical Small-Signal Approximations:

g_m >> g_π, g_{mb} >> g_o, g_{ds} ∴ $\overbrace{\beta_0}$

r_{gm} << r_π, r_{gmb} << r_o, r_{ds} << $(g_m r_\pi) r_o, (g_m r_{ds}) r_{ds}$

C_μ, C_{GD} << C_π, C_{GS} << $C_{INTENTIONAL}$ << $C_{OFF\text{-}CHIP}$

Low   << Moderate <<   High   <<   Very High $C_{LD} \geq C_{\pi/GS}$

B. Signal Flow

Possible Paths

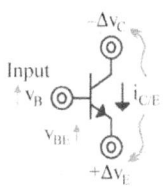

B/G Input: Higher $v_{B/G}$ raises $v_{BE/GS}$

   ∴   Raises $i_{C/E}$ → Reduces v_C Raises v_E

   → $v_{B/G} : v_{E/S}$ Translation = In phase

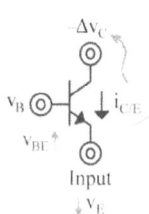

        $v_{B/G} : v_{C/D}$ Translation = Out of phase

E/S Input: Lower $v_{E/S}$ raises $v_{BE/GS}$

   ∴   Raises $i_{C/E}$ → Reduces v_C

   → $v_{E/S} : v_{C/D}$ Translation = In phase

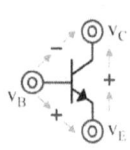

Conclusion: Only CE/S stage inverts

C. Trans–Ohmic Translations: i. Accurate Expressions

Transconductance to $i_{g/e/s}$:

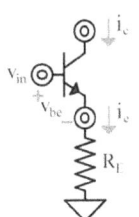
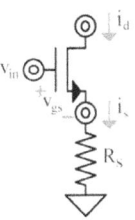

$v_{be/gs} = v_{in}$ fraction

→ $i_{e/s}R_{E/S}$ degenerates $v_{be/gs}$

$i_{e/s} = i_\pi + i_g + i_{gb} + i_{ro/ds}$

i_π feeds v_e, v_e feeds $i_{ro} = \dfrac{i_\pi R_E}{r_o}$ to $v_c = 0$

$$G_M \equiv \left.\frac{i_{g/e/s}}{v_{b/g}}\right|_{v_{c/d}=0} = \frac{\dfrac{1}{r_\pi} \pm g_m + \left(\dfrac{1}{r_\pi}\right)\left(\dfrac{R_E}{r_o}\right)}{1 + \left(\dfrac{1}{r_\pi} + g_m + g_{mb} + \dfrac{1}{r_{o/ds}}\right)R_{E/S}}$$

↓

G_M to i_e: i_g & i_π feed $v_{be/gs}g_m$ & $\dfrac{v_{be}}{r_\pi}$ to v_e

2-port G_M model to i_g for CE/S

BJT excludes g_{mb} MOS excludes r_π

Direct translation to $i_{e/s}$ for CC/D

Base/Collector/Drain Resistance :

$R_{E/S}$ loads $R_{B(EQ)}$ & $R_{C/D(EQ)}$

$i_{e/s}R_{E/S}$ degenerates $v_{be/gs}$

$r_\pi + R_B$ parallels R_E

$v_\pi = v_e$ fraction

→ $i_\pi R_B$ degenerates v_π

$v_{gs}g_m + v_{bs}g_{mb} = v_g g_m + v_b g_{mb} - v_s(g_m + g_{mb})$

→ g_{mb} reinforces v_s degeneration of g_m

When $i_g \gg i_{ro}$

↓

$$R_{B(EQ)} \approx r_\pi + R_E + \overbrace{g_m r_\pi}^{\beta_0} R_E$$

$$R_{C/D(EQ)} = r_{o/ds} + R_{E/S}' + (g_m' + g_{mb})r_{o/ds}R_{E/S}'$$

$$R_E' = (r_\pi + R_B) \| R_E$$

$$v_\pi g_m = v_e \overbrace{\left(\frac{r_\pi}{r_\pi + R_B}\right)g_m}^{g_m'}$$

Emitter/Source Resistance:

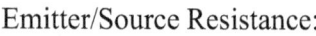

$$r_\pi \| \frac{1}{g_m} \| r_o = r_\pi \| \frac{r_o}{1 + g_m r_o}$$

R_B loads r_π

$R_{C/D}$ loads $r_{gm} \| r_{o/ds}$

$$R_{E/S(EQ)} = (r_\pi + R_B) \| \left[\overbrace{\frac{r_{o/ds} + R_{C/D}}{1 + (g_m{}' + g_{mb}) r_{o/ds}}}^{R_{E/SG}} \right]$$

$i_\pi R_B$ degenerates v_π

$v_{bs} g_{mb}$ reinforces $v_{gs} g_m$

$$v_\pi g_m = v_{in} \left(\frac{r_\pi}{r_\pi + R_B} \right) g_m \equiv v_{in} g_m{}'$$

Plain $R_{E/SG}$: $\dfrac{1}{g_m + g_{mb}} \| r_{o/ds}$ Base-degenerated R_{EG}: $\dfrac{1}{g_m{}'} \| r_o$

Loaded $R_{E/SG}$: $\dfrac{r_{o/ds} + R_{C/D}}{1 + (g_m + g_{mb}) r_{o/ds}}$

$$\left[\begin{array}{l} \text{If } R_{C/D} \approx r_{o/ds} \;\rightarrow\; R_{E/SG} \approx \dfrac{2}{g_m} \\[2em] \text{If } R_{C/D} \approx (g_m r_{o/ds}) r_{o/ds} \;\rightarrow\; R_{E/SG} \approx r_{o/ds} \end{array} \right.$$

ii. Translations

Analysis: Static signals do not vary $\rightarrow$ $s_X = S_X + s_x = S_X + 0$ $\rightarrow$ $s_x = 0$

           Trace small-signal path

           Track $v_x : i_y$ & $i_y : v_z$ translations

Example: Transconductor M_{CS} : Current Buffer Q_{CB} : Voltage Follower Q_{CC}

           $i_{in} \rightarrow v_{in} \rightarrow i_{gCS} \rightarrow v_{eCB} \rightarrow i_{cCB} \rightarrow v_{o1} \rightarrow i_{eCC} \rightarrow v_o$

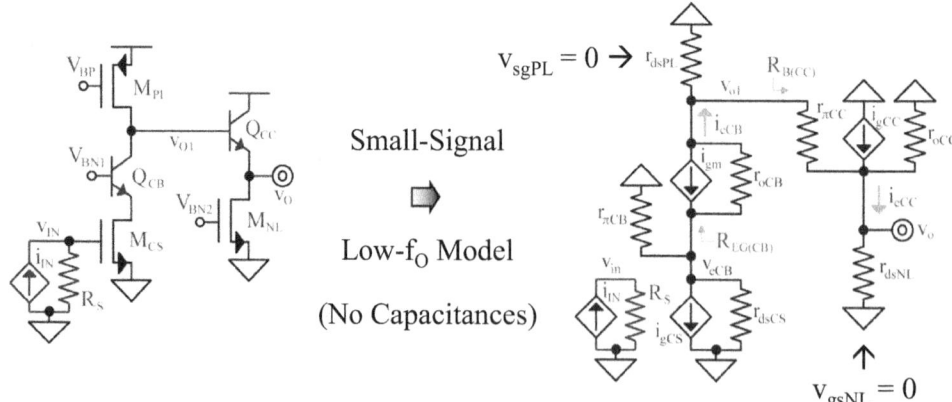

Small-Signal

➡️

Low-f_O Model

(No Capacitances)

$$\frac{v_o}{i_{in}} = \left(\frac{v_{in}}{i_{in}}\right)\left(\frac{v_{ol}}{v_{in}}\right)\left(\frac{v_o}{v_{ol}}\right) = A_Z A_{V1} A_{V2} = R_S\left[\left(\frac{i_{gCS}}{v_{in}}\right)\left(\frac{v_{eCB}}{i_{gCS}}\right)\left(\frac{i_{cCB}}{v_{eCB}}\right)\left(\frac{v_{ol}}{i_{cCB}}\right)\right]\left[\left(\frac{i_{eCC}}{v_{ol}}\right)\left(\frac{v_o}{i_{eCC}}\right)\right]$$

$$A_{GCS} \approx -g_m \qquad A_{ICB} \approx 1 \quad \approx r_{dsPL} \qquad A_{VCC} \approx 1$$

$$A_{V1} = (-g_{mCS})(r_{dsCS} \parallel r_{\pi CB} \parallel R_{EG(CB)})\left(\frac{1}{R_{EG(CB)}}\right)(r_{dsPL} \parallel R_{B(CC)}) \approx -g_{mCS}r_{dsPL}$$

$$A_{V2} = \left[\frac{\dfrac{1}{r_{\pi CC}} + g_{mCC}}{1 + \left(\dfrac{1}{r_{\pi CC}} + g_{mCC} + \dfrac{1}{r_{oCC}}\right)r_{dsNL}}\right] r_{dsNL} \approx 1$$

Where: $R_{EG(CB)} = \dfrac{r_{oCB} + (r_{dsPL} \parallel R_{B(CC)})}{1 + g_{mCB}r_{oCB}} \approx \dfrac{r_{oCB} + r_{dsPL}}{g_{mCB}r_{oCB}} \rightarrow \dfrac{2}{g_{mCB}}$

$$R_{B(CC)} = r_{\pi CC} + (r_{dsNL} \parallel r_{oCC}) + \underbrace{g_{mCC}r_{\pi CC}(r_{dsNL} \parallel r_{oCC})}_{\beta_0}$$

1.4. Frequency Response: A. Shunt Capacitor

Gain: $A_V \equiv \dfrac{v_O}{v_{IN}} = \dfrac{R_O \parallel Z_S}{R_C + (R_O \parallel Z_S)} = \dfrac{R_C \parallel R_O \parallel Z_S}{R_C} = \dfrac{A_{V0}}{1 + s/2\pi p_C}$

C_S opens at low f_O: Zero/Low-f_O Gain $A_{V0} \equiv A_V\big|_{f_O \to 0} = \dfrac{R_O}{R_C + R_O}$

C_S shunts R_O (R_O fades) when $Z_S = \dfrac{1}{sC_S} < R_O$ past $f_{RO} = \dfrac{1}{2\pi R_O C_S}$

C_S shunts parallel $R = R_C \parallel R_O$ past $p_C = \dfrac{1}{2\pi(R_C \parallel R_O)C_S}$

C_S shorts with f_O Capacitor Pole $\leftarrow$ –20 dB per decade

High-f_O Gain $A_{V(HF)} \equiv A_V\big|_{f_O \to \infty} = \dfrac{Z_S}{R_C} \propto \dfrac{1}{sR_C C_S} \propto \dfrac{1}{f_O}$ Falls 10× with 10× rise in f_O

C_S delays i_C–v_C sinusoids: Phase shifts up to –90° past p_C $\rightarrow$ $\angle A_V = -\tan^{-1}\left(\dfrac{f_O}{p_C}\right)$

Current-Limit Resistor

Gain: $A_V = \dfrac{R_O \| (R_I + Z_S)}{R_C + \left[R_O \| (R_I + Z_S) \right]} = \dfrac{R_C \| R_O \| Z_S'}{R_C} = A_{V0}\left(\dfrac{1 + s/2\pi z_{CX}}{1 + s/2\pi p_C} \right)$

C_S & R_I shunt parallel $R = R_C \| R_O$ when $Z_S < R_I + (R_C \| R_O)$

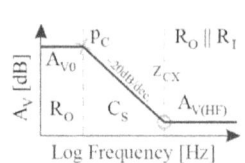

past $p_C = \dfrac{1}{2\pi \left[R_I + (R_C \| R_O) \right] C_S}$

$\qquad$ Reversal Zero

R_I limits $i_S = \dfrac{v_O}{R_I + \cancelto{0}{Z_S}}$ when $Z_S = \dfrac{1}{sC_S} < R_I$ past $z_{CX} = \dfrac{1}{2\pi R_I C_S}$

→ C_S effects fade (C_S shorts) → p_C reverses → $p_C < z_{CX}$

$A_{V0} = \dfrac{R_O}{R_C + R_O}$ $\qquad$ $A_{V(HF)} = \dfrac{R_O \| R_I}{R_C + (R_O \| R_I)}$

Phase: z_{CX} recovers up to 90° $\qquad$ → $\qquad$ $\angle A_V = -\tan^{-1}\left(\dfrac{f_O}{p_C} \right) + \tan^{-1}\left(\dfrac{f_O}{z_{CX}} \right) \geq -90°$

B. Cross-Amp (Bypass) Capacitor: i. Modeling

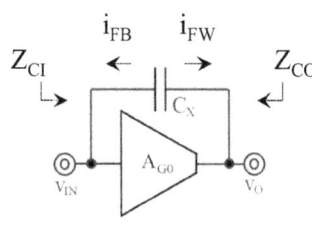 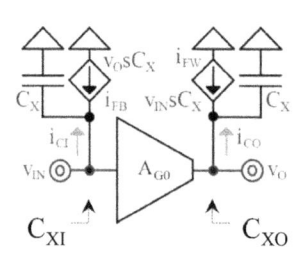

FB : FW Model: $\quad Z_{CI}$ when $v_O = 0$ → $Z_C = \dfrac{1}{sC_X}$ ← Z_{CO} when $v_{IN} = 0$

i_{FB} when $v_{IN} = 0$ → $\dfrac{v_O}{Z_C} = v_O sC_X$ $\qquad$ i_{FW} when $v_O = 0$ → $\dfrac{v_{IN}}{Z_C} = v_{IN} sC_X$

$C_{XI} : C_{XO}$ Model: $\quad G_{CI} \equiv \dfrac{i_{CI}}{v_{IN}} = \dfrac{1}{Z_C} - \dfrac{v_O sC_X}{v_{IN}} = (1 - A_V)sC_X = sC_{XI}$

$\qquad\qquad\qquad G_{CO} \equiv \dfrac{i_{CO}}{v_O} = \dfrac{1}{Z_C} - \dfrac{v_{IN} sC_X}{v_O} = \left(1 - \dfrac{1}{A_V} \right)sC_X = sC_{XO}$

ii. In-Phase Capacitor: Across Non-Inverting Amp

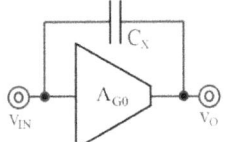

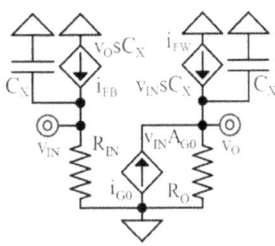

Buffer: $A_V \approx +1$ V/V

$$C_{XI} = (1 - A_V)C_X \approx 0 \qquad C_{XO} = \left(1 - \frac{1}{A_V}\right)C_X \approx 0 \qquad \rightarrow \qquad \text{Buffer removes } C_{XI}, C_{XO}$$

Gain: Z_C falls with f_O $\rightarrow$ i_C rises with f_O $\therefore$ i_{FW} overcomes i_{G0} past z_C

$$A_G \equiv \left.\frac{i_G}{v_{IN}}\right|_{v_O=0} \equiv \left.\frac{i_{G0} + i_{FW}}{v_{IN}}\right|_{v_O=0} = A_{G0} + sC_X = A_{G0}\left(1 + \frac{sC_X}{A_{G0}}\right) = A_{G0}\left(1 + \frac{s}{2\pi z_C}\right)$$

i_{FW} replenishes C_O: Recovers up to 90° $\rightarrow$ $\angle A_G = +\tan^{-1}\left(\frac{f_O}{z_C}\right)$ $\quad i_{FW}$ reinforces i_{G0}

In-phase zero

ii. Out-of-Phase Capacitor: Across Inverting Amp

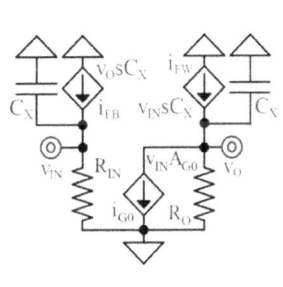

Inv. Amp: $C_{XI} = (1 - A_V)C_X \geq C_X$ $\rightarrow$ A_V increases C_{XI}

$A_V < 0$ V/V $\quad C_{XO} = \left(1 - \frac{1}{A_V}\right)C_X \approx C_X$ when $|A_V| \gg 1$ V/V

Gain: Z_C falls with f_O $\rightarrow$ i_C rises with f_O $\therefore$ i_{FW} overcomes i_{G0} past z_C

$$A_G \equiv \left.\frac{i_G}{v_{IN}}\right|_{v_O=0} \equiv \left.\frac{i_{G0} - i_{FW}}{v_{IN}}\right|_{v_O=0} = A_{G0} - sC_X = A_{G0}\left(1 - \frac{sC_X}{A_{G0}}\right) = A_{G0}\left(1 - \frac{s}{2\pi z_C}\right)$$

i_{FW} opposes i_{G0} $\rightarrow$ Inverts i_G $\rightarrow$ Inverting out-of-phase zero $\leftarrow$

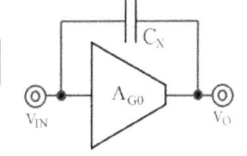

i_{FW} replenishes C_O, inverts i_G: Up to $+90° - 180° = -90°$ $\rightarrow$ $\angle A_G = -\tan^{-1}\left(\frac{f_O}{z_C}\right)$

15

C. Recursive Shunt-Circuit Transformations

Low f_O : $\quad$ C's = Open at low f_O $\quad \therefore$ $\quad$ Derive A_{X0} without C's

f_O Response: $\quad$ Analyze shunt-RC network $\quad \therefore$ $\quad$ Split (coupling) cross-amp C's

$\quad$ Open-circuit (OC) approximation $\quad \rightarrow$ $\quad$ Highest RC shunts first

$\quad$ C_{EQ1} shunts parallel R_{EQ1} first past $p_1 \approx f_1 = \dfrac{1}{2\pi R_{EQ1}C_{EQ1}}$

$\quad$ Shunt-capacitor (SC) approximation $\quad \rightarrow$ $\quad$ R_{EQ1} fades past p_1

$\quad$ Remove R_{EQ1}, merge C's, split cross-amp C's, use OC to find p_2

$\quad$ C_{EQ2} shunts parallel R_{EQ2} second past $p_2 \approx f_2 = \dfrac{1}{2\pi R_{EQ2}C_{EQ2}}$

$\quad$ Repeat for every shunt-RC node $\quad \therefore$ $\quad$ K shunt nodes $\rightarrow$ K poles

Clusters: $\quad$ N poles are within a decade of one another

Coupled: $\quad$ Cross-amp C's couple poles when cross-amp z_X's are in cluster

$\therefore$ $\quad$ $p_1 \approx \dfrac{1}{2\pi\Sigma R_N C_N} = f_1 \parallel ... \parallel f_N$ $\qquad$ $p_2 \approx p_1\left(\dfrac{f_2}{f_1}\right)$ $\qquad ... \qquad$ $p_N \approx p_1\left(\dfrac{f_N}{f_1}\right)$

Couplers: $\quad$ $C_{\mu/GD}$ in transconductors & $C_{\pi/GS}$ in followers

Uncoupled: $\quad$ Apply same SC/OC approximation to clustered poles

$\therefore$ $\quad$ $p_1 \approx f_1 = \dfrac{1}{2\pi R_1 C_1}$ $\qquad$ $p_2 \approx f_2 = \dfrac{1}{2\pi R_2 C_2}$ $\qquad ... \qquad$ $p_N \approx f_N = \dfrac{1}{2\pi R_N C_N}$

Decouplers: $\quad$ Current buffers

Note: $\quad$ Short-/open-circuit R's can alter/distort RC net $\rightarrow$ No 2-port R's

D. Gain–Bandwidth Product ≡ GBW

Condition: Drop in gain past 1 pole

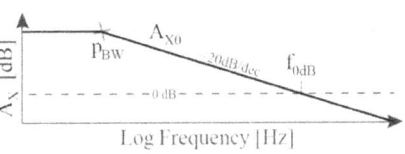

$A_{X0} \equiv$ Low-f_O gain

$f_{0dB} \equiv$ Unity-gain frequency

$p_{BW} \equiv$ Bandwidth-Setting Pole $\equiv$ –3-dB f_O

Past p_{BW}: $GBW \equiv A_{X0}p_{BW} \approx A_X f_O\big|_{f_O \geq p_{BW}} = (1)f_{0dB} = $ Constant

A_X falls 10× every 10× rise in f_O → Fall in A_X cancels rise in f_O

∴ Use GBW to project A_X past $p_{BW} \approx \dfrac{GBW}{f_O}$ & $f_{0dB} \approx GBW$

E. CS–CD Example

Low-f_O Gains: $A_{V10} = -g_{m1}(r_{ds1} \parallel r_{ds2})$ $A_{V40} \approx 1$ V/V

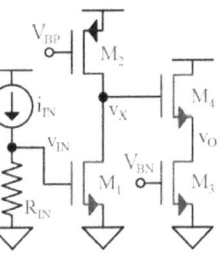

$A_{Z0} \equiv \dfrac{v_o}{i_{in}} = R_{IN}A_{V10}A_{V40} \approx R_{IN}A_{V10}$

p_1 OC Model: Split C_{GD1}, C_{GS4}

→ A_{V10} increases C_{GD1I} A_{V40} removes C_{GS4}

$R_{IN}'C_{IN}' = R_{IN}[C_{GS1} + (1 - A_{V10})C_{GD1}]$

$R_X'C_X' \approx (r_{ds1} \parallel r_{ds2})(C_{GD1} + C_{GD2} + C_{GD4})$

$R_O'C_O' \approx \left(\dfrac{1}{g_{m4}} \parallel r_{ds3} \parallel r_{ds4}\right)C_{GD3}$ If $f_{IN}' \ll f_X', f_O'$ ∴ $p_1 \approx f_{IN}' = \dfrac{1}{2\pi R_{IN}'C_{IN}'}$

p_2 SC/OC Model: R_{IN} fades past p_1

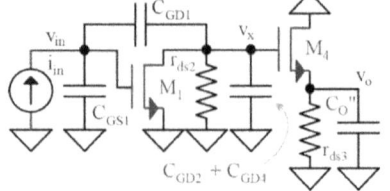

Split C_{GS4} → No C_{GS4}

Shunt R's: $R_X"$, $R_O"$ → 2 poles

Analysis: Split C_{GD1}'s i_{FW} : i_{FB} effects

FW: i_{fw1} overcomes i_{g1} past inverting $z_{GD1} = \dfrac{g_{m1}}{2\pi C_{GD1}}$

FB: C_{GD1}:C_{GS1} voltage-divides v_x to v_{in}, i_{g1} shunts v_x → R_{G1} shunts $R_X"$

$$R_X"C_X" \approx \left(\frac{C_{GD1}+C_{GS1}}{C_{GD1}g_{m1}} \parallel r_{ds1} \parallel r_{ds2} \right)[(C_{GD1} \oplus C_{GS1}) + C_{GD2} + C_{GD4}]$$

$$R_O"C_O" \approx \left(\frac{1}{g_{m4}} \parallel r_{ds3} \parallel r_{ds4} \right) C_{GD3} \qquad \text{If } f_X" << f_O" \quad \therefore \quad p_2 \approx f_X" = \frac{1}{2\pi R_X"C_X"}$$

p_3 SC/OC Model: R_{IN}, r_{ds1}, r_{ds2} fade past p_2

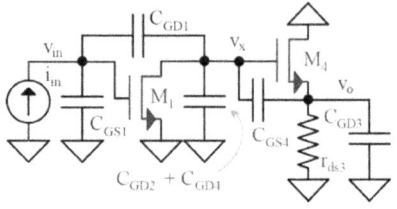

Shunt R's: R_O''' → 1 pole

Analysis: Split C_{GS4}'s i_{FW} : i_{FB} effects

FW: i_{fw4} overcomes i_{g4} past in-phase $z_{GS4} = \dfrac{g_{m4}}{2\pi C_{GS4}}$

FB: C_{GS4}:C_{EQ} voltage-divides v_o to v_{sg4}, i_{g4} shunts v_o → R_{G4} shunts R_O'''

$$R_O'''C_O''' \approx \left(\frac{C_{GS4}+C_{EQ}}{C_{EQ}g_{m4}} \parallel r_{ds3} \parallel r_{ds4} \right)[(C_{GS4} \oplus C_{EQ}) + C_{GD3}]$$

$$C_{EQ} = (C_{GD1} \oplus C_{GS1}) + C_{GD2} + C_{GD4} \qquad \therefore \qquad p_3 \approx f_O''' = \frac{1}{2\pi R_O'''C_O'''}$$

Chapter 2. Analog Primitives

2.1. Cascode

2.2. Current Mirror

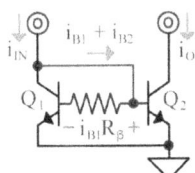

2.3. E/S-Coupled Pair

2.4. B/G-Coupled Pair

2.5. Simulations

2.1. Cascode

Origin: Triode ≡ 3-Terminal Diode   Cascode ≡ Cascaded Triodes

Low-f_O Gains: $A_{V10} = -g_{m1}(r_{ds1} \parallel R_{S2})$ → $-g_{m1}\left(\dfrac{2}{g_{m2}}\right)$ →   Low

$$A_{Z0} \equiv \frac{v_o}{i_{in}} = R_{IN}A_{V10}\left(\frac{1}{R_{S2}}\right)R_{LD} \approx -R_{IN}g_{m1}R_{LD}$$

→   Adding M_2 reduces A_{V10} without altering A_{Z0} much

p_1 OC Model: Split C_{GD1} →   Low A_{V10} keeps C_{GD11} low   ∴   Higher f_{IN}'

Shunt R's: R_{IN}', R_X', R_O' →   3 poles   ∴ M_2 adds p_X

p_X: $R_X' = r_{ds1} \parallel R_{S2} \approx R_{S2}$ →   Low   →   High p_X ∴ M_2 raises p_X

2.2. Current Mirror: A. Large Signal

Diode Connection

Operation:

Short/bypass BC diode → BE diode

$i_{C/D}$ = Sensitive to $v_{BE/GS}$

= Insensitive to $v_{CE/DS}$ in Act./Sat.

When $v_{BE/GS}$'s match   ∴ $i_O \approx i_{IN}$

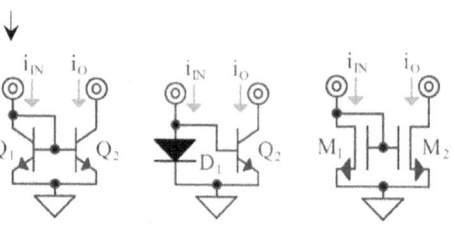

Translations:

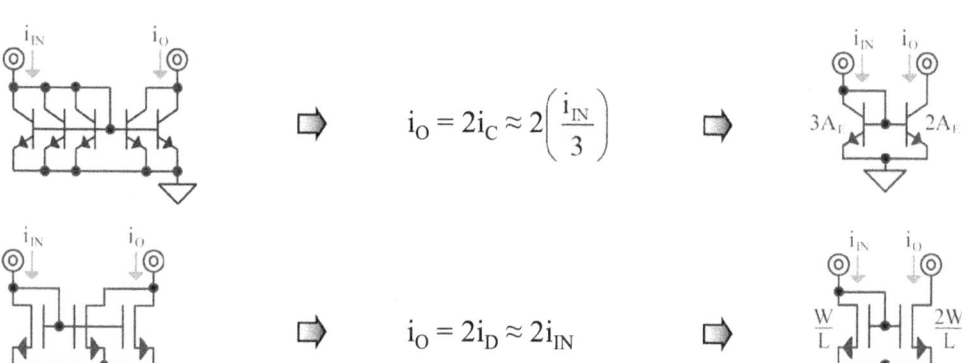

$$i_O = 2i_C \approx 2\left(\frac{i_{IN}}{3}\right)$$

$$i_O = 2i_D \approx 2i_{IN}$$

$v_{C/D}$ Mismatch: $v_{C/D1} \neq v_{C/D2}$

Error

$$A_{IC} \equiv \frac{i_{C2}}{i_{C1}} = \frac{I_{S2}\left[\exp\left(\frac{v_{BE}}{V_t}\right)-1\right]\left(1+\frac{v_{CE2}}{V_A}\right)}{I_{S1}\left[\exp\left(\frac{v_{BE}}{V_t}\right)-1\right]\left(1+\frac{v_{CE1}}{V_A}\right)} = \left(\frac{A_{E2}}{A_{E1}}\right)\left(\frac{1+\frac{v_O}{V_A}}{1+\frac{v_{BE}}{V_A}}\right)$$

$$A_{ID} \equiv \frac{i_{D2}}{i_{D1}} = \frac{\left(\frac{W}{L}\right)_2 K'(v_{GS}-v_T)^2(1+\lambda v_{DS2})}{\left(\frac{W}{L}\right)_1 K'(v_{GS}-v_T)^2(1+\lambda v_{DS1})} = \left(\frac{S_2}{S_1}\right)\left(\frac{1+\lambda v_O}{1+\lambda v_{GS}}\right)$$

Error

i_B Error: $i_O = A_{IC}i_{C1} < A_{IC}i_{IN} = A_{IC}(i_{C1}+i_{B1}+i_{B2})$

$$A_{IO} \equiv \frac{i_O}{i_{IN}} = \frac{i_{C2}}{i_{C1}+i_{B1}+i_{B2}} = \frac{A_{IC}i_{C1}}{i_{C1}+\frac{i_{C1}}{\beta_0}+\frac{A_{IC}i_{C1}}{\beta_0}} = \frac{A_{IC}}{1+\frac{1}{\beta_0}+\frac{A_{IC}}{\beta_0}} < A_{IC} \qquad S_i \equiv \frac{W_i}{L_i}$$

B. Error Correction: i. $v_{C/D}$

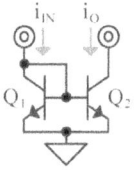

Mismatch: $v_O \neq v_{IN}$ → $v_{CE/DS2} \neq v_{CE/DS1}$

∴ $A_{IC/D} = f(v_{CE/DS}\text{'s})$

Fix: Match $v_{CE/DS}$'s → Cascodes $v_O \geq v_{CE2(MIN)}$

| Basic: | Q$_2$ mirrors Q$_1$ | Low Voltage: | | Self-Biased: |

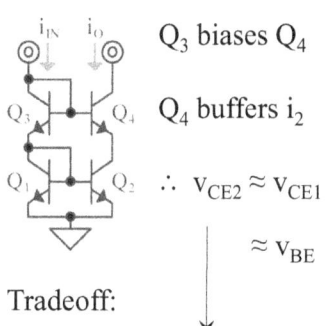

Basic: Q_2 mirrors Q_1

Q_3 biases Q_4

Q_4 buffers i_2

∴ $v_{CE2} \approx v_{CE1}$

$\approx v_{BE}$

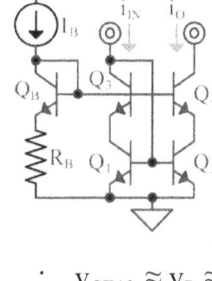

Low Voltage:

I_B:Q_B:R_B bias Q_{34}

Q_3 buffers i_1

Q_3 mirror-

-connects Q_{12}

∴ $v_{CE12} \approx v_R \approx I_B R_B \neq f(i_{IN})$

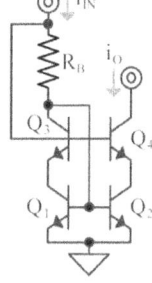

Self-Biased:

$v_{CE12} \approx v_R \approx i_{IN} R_B$

Tradeoff:

$v_O \geq v_{CE4(MIN)} + v_{BE}$ $v_O \geq v_{CE4(MIN)} + v_R$

Design Aim: $v_R \geq v_{CE(MIN)}$ $v_{DS(SAT)}$

Across i_{IN}, T_J, & fabrication corners

MOS Variation: v_T's match without body effect

$$v_{DS12} = v_{GSB} - v_{GS34} \approx \underbrace{\sqrt{\frac{2I_B}{(W/L)_B K_N'}}}_{v_{DSB(SAT)}} - \underbrace{\sqrt{\frac{2i_{IN}}{(W/L)_3 K_N'}}}_{v_{DS34(SAT)}} \geq v_{DS12(SAT)} \approx \sqrt{\frac{2i_{IN}}{(W/L)_1 K_N'}}$$

— Design aim

→ Same v_T & K_N' effects, but dissimilar to i_{IN} effects in $v_{DS12(SAT)}$

With body effect: $v_{DS12} \approx \sqrt{\frac{2I_B}{(W/L)_B K_N'}} - \sqrt{\frac{2i_{IN}}{(W/L)_3 K_N'}} - \gamma_N \left(\sqrt{2\psi_B + \underset{v_{SB4}}{v_{DS12}}} - \sqrt{2\psi_B} \right)$

→ Dissimilar to i_{IN} & v_T effects in $v_{DS12(SAT)}$

ii. i_B Small negative error $\rightarrow$ Small-signal effect

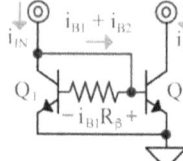

Error: $i_O = i_{C1}A_{IC} = (i_{IN} - i_{B1} - i_{B2})A_{IC} \neq i_{IN}A_{IC}$

Compensate: Raise i_{C2} $\rightarrow$ Raise v_{BE2} $\rightarrow$ $v_{BE2} = v_{BE1} + i_{B1}R_\beta$

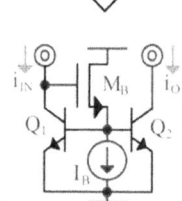

$\Delta i_O = i_{IN}A_I - i_{C1}A_{IC} \equiv v_R g_{m2} = i_{B1}R_\beta g_{m2}$ $R_\beta = \left(\dfrac{\beta_0}{i_{C1}}\right)\left(\dfrac{\Delta i_O}{g_{m2}}\right)$

Reduce i_E: β_0-suppress i_E $\rightarrow$ Q_B mirror-connects Q_{12}

$$I_B \text{ raises } p_B \approx \frac{1}{2\pi R_B C_B} \approx \frac{g_{mB}'}{2\pi C_B} \propto I_B$$

$$I_B \text{ raises } i_E = \frac{I_B + \dfrac{\overbrace{i_O/A_{IC}}^{i_{C1}}}{\beta_0} + \dfrac{\overbrace{i_O}^{i_{C2}}}{\beta_0}}{1+\beta_0} \approx \frac{I_B}{\beta_0} + \left(\frac{i_O}{\beta_0^2}\right)\left(1 + \frac{1}{A_{IC}}\right)$$

Eliminate i_E: M_B mirror-connects Q_{12} $\rightarrow$ M_B (if available) blocks i_E

C. Small-Signal Gain

Diode Connection:

     v_{BC} connection shorts & disables base–collector PN junction

     $\therefore$ $Q_1 = $ BE Diode $\rightarrow$ $R_D = r_o \parallel \dfrac{1}{g_m} \parallel r_\pi \approx \dfrac{1}{g_m}$

Current Mirror:

Q_1 & $M_1 = $ Diode-Connected

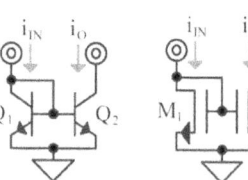

$$A_{I0} \equiv \frac{i_o}{i_{in}} = \left(\frac{v_{in}}{i_{in}}\right)\left(\frac{i_{g2}}{v_{in}}\right)\left(\frac{v_o}{i_{g2}}\right)\left(\frac{i_o}{v_o}\right)$$

$$= \left(R_S \parallel r_{o1} \parallel \frac{1}{g_{m1}} \parallel r_{\pi1} \parallel r_{\pi2}\right)(-g_{m2})(r_{o2} \parallel R_{LD})\left(\frac{-1}{R_{LD}}\right)$$

$$\approx \frac{g_{m2}}{g_{m1}} \approx \left(\frac{I_2}{V_t}\right)\left(\frac{V_t}{I_1}\right) \approx \frac{A_{E2}}{A_{E1}} \approx A_{IC} \qquad \text{If} \qquad R_S \gg R_D \qquad R_{LD} \ll r_{o2}$$

2.3. E/S-Coupled Pair: A. Large Signal

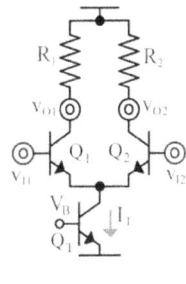

Inputs: Common $v_{IC} \equiv avg = \dfrac{v_{I1} + v_{I2}}{2}$

   Differential $v_{ID} \equiv v_{I1} - v_{I2} = \left(\dfrac{+v_{ID}}{2} \right) - \left(\dfrac{-v_{ID}}{2} \right)$

Bias: Q_{12} amplifies v_{ID} $\therefore$ $v_{ID} = \dfrac{v_{O12}}{A_{D12}}$ $\rightarrow$ Low

   $\rightarrow$ $V_{I1} \approx V_{I2}$ $\rightarrow$ $V_{BE1} \approx V_{BE2}$ $\rightarrow$ $I_{C1} \approx I_{C2} \approx 0.5I_T$

Input Common-Mode Range (ICMR):

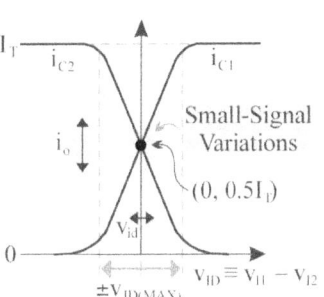

   Low v_{IC} "crushes" Q_T into deep sat.

   $\therefore$ $v_{IC} \geq v_{EE} + v_{CET(MIN)} + v_{BE12}$

   High v_{IC} "crushes" Q_{12} into R_{12} load

   $\therefore$ $v_{IC} \leq v_{CC} - v_{LD12} - v_{CE1(MIN)} + v_{BE12}$

B. Small v_{ID}

Small Signals $\therefore$ Exclude static components $\rightarrow$ V_{ID} V_{IC} V_{CC} V_{EE} I_T

Superposition $\therefore$ Exclude other small signals $\rightarrow$ $v_{ic} \equiv v_{cc} \equiv v_{ee} \equiv i_t \equiv 0$

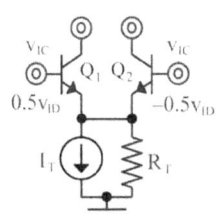

Parameters: $R_{ID} \equiv \dfrac{v_{id}}{i_{id}} = \dfrac{v_{id}}{0.5v_{id}/r_{\pi 12}} = 2r_{\pi 12}$ $R_{OD12} \equiv \left. \dfrac{v_{o12}}{i_{o12}} \right|_{v_{id}=0} = r_{o12}$

   $G_{D1} \equiv \left. \dfrac{i_{o1}}{v_{id}} \right|_{v_{o1}=0} = -0.5g_{m12}$ $G_{D2} \equiv \left. \dfrac{i_{o2}}{v_{id}} \right|_{v_{o1}=0} = +0.5g_{m12}$

C. Small v_{IC}

Small Signals $\therefore$ Exclude static components $\rightarrow$ V_{ID} V_{IC} V_{CC} V_{EE} I_T

Superposition $\therefore$ Exclude other small signals $\rightarrow$ $v_{id} \equiv v_{cc} \equiv v_{ee} \equiv i_t \equiv 0$

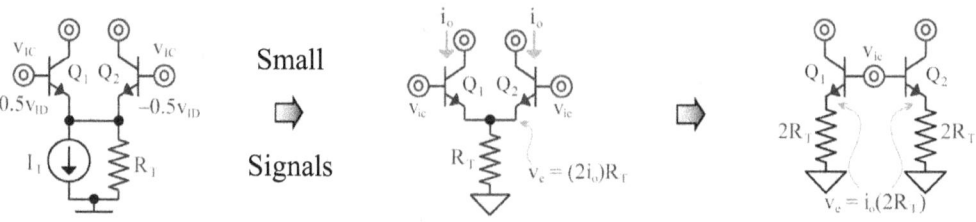

Parameters:

$$R_{IC12} \equiv \frac{V_{ic}}{i_{ic12}} = R_{B12} \approx r_{\pi12} + 2R_T + g_{m12}r_{\pi12}(2R_T)$$

$$R_{OC12} \equiv \frac{V_{o12}}{i_{o12}}\bigg|_{v_{ic}=0} = R_{C12} = r_{o12} + \left(2R_1 \| r_{\pi12}\right) + g_{m12}r_{o12}\left(2R_T \| r_{\pi12}\right)$$

$$G_{C12} \equiv \frac{i_{o12}}{V_{ic}}\bigg|_{v_{o12}=0} = \frac{\left(1/r_{\pi12}\right)\left(2R_T/r_{o12}\right) - g_{m12}}{1 + \left(1/r_{\pi12} + g_{m12} + 1/r_{o12}\right)\left(2R_T\right)}$$

D. Differential Range

$$\pm V_{ID(MAX)} \equiv \pm \frac{\Delta i_{O12(MAX)}}{G_{D12}} = \pm \frac{I_T - 0.5I_T}{G_{D12}} \approx \pm \frac{0.5I_T}{0.5g_{m12}} = \pm \frac{I_T}{g_{m12}}$$

Exponential: BJT   Sub-v_T MOS

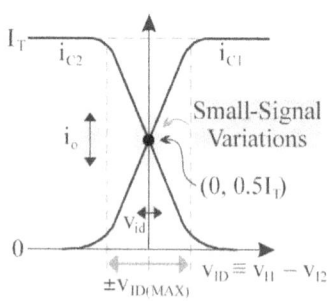

$$\pm V_{ID(MAX)} \approx \pm I_T \left(\frac{n_1 V_t}{0.5I_T}\right) = \pm 2n_1 V_t$$

Square Law: JFET   Inverted MOS

$$\pm V_{ID(MAX)} \approx \pm \frac{I_T}{\sqrt{2\left(0.5I_T\right)K'\left(\dfrac{W}{L}\right)}} = \pm \sqrt{\frac{2\left(0.5I_T\right)}{K'\left(\dfrac{W}{L}\right)}} = \pm V_{DS(SAT)}$$

E. Degeneration

$$v_e' = (i_o - i_o)R_T = 0$$

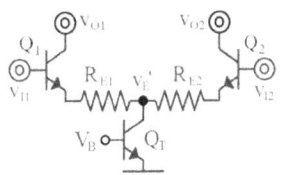 Small ⇨ Signals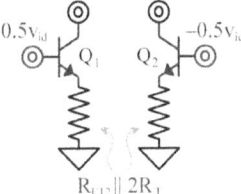

$0.5I_TR_{E12}$ raises $v_{IC(MIN)}$ R_{E12} reduces G_D → Extends $v_{ID(MAX)}$

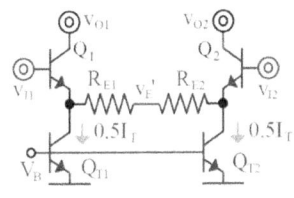 Small  Signals

$v_{I1} \approx v_{I2}$ $v_{E1} \approx v_{E2}$ ∴ $i_{RE} \approx 0$ →   No effect in $v_{IC(MIN)}$ or G_C

Requirement: $Q_{T1}:Q_{T2}$ mismatch produces imbalance → $Q_{T1}:Q_{T2}$ should match

E. MOS

ICMR: $v_{IC} \geq v_{SS} + v_{DST(SAT)} + v_{GS12} = v_{SS} + v_{DST(SAT)} + v_{TN} + v_{DS12(SAT)}$

$v_{IC} \leq v_{DD} - v_{LD12} - v_{DS12(SAT)} + v_{GS12} \leq v_{DD} - v_{LD12} + v_{TN12}$

Body: To common source   ∴ C_{SUB} couples substrate noise to v_S

To substrate = v_{SS} ∴ C_{BS} couples v_{SS} noise to M_{12} channels

Body effect shifts v_T → ICMR shifts

Variants:

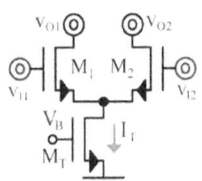 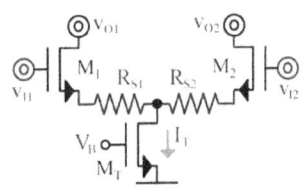 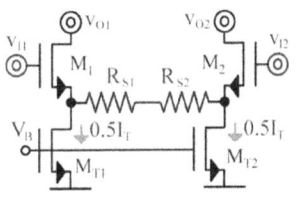

2.4. B/G-Coupled Pair: A. Large Signal

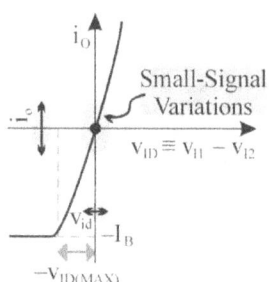

Inputs: Common $v_{IC} \equiv \text{avg} = \dfrac{v_{I1} + v_{I2}}{2}$

Differential $v_{ID} \equiv v_{I1} - v_{I2} \equiv \left(\dfrac{+v_{ID}}{2}\right) - \left(\dfrac{-v_{ID}}{2}\right)$

Bias: Q_{12} amplifies v_{ID} $\therefore$ $v_{ID} = \dfrac{v_O}{A_D}$ $\rightarrow$ Low

$\therefore$ $V_{I1} \approx V_{I2}$ $\rightarrow$ $V_{EB1} \approx V_{EB2}$ $\rightarrow$ $I_{C1} \approx I_{C2} \approx I_B$ $\rightarrow$ $I_O \approx 0$

v_{IC} Range:

Low v_{IC} "crushes" Q_2 into Q_{B2}

$\rightarrow$ $v_{IC} \geq v_{EE} + v_{CEB2(MIN)} + v_{EB2}$

Input or v_{CE} breakdown limits $v_{IC(MAX)}$

i_O Range: $i_{O(MIN)} = -I_B$ $\qquad$ $i_{O(MAX)} \approx i_{I1(MAX)} - I_B$ $\rightarrow$ v_{I1} limits i_{I1}

B. Small v_{ID}

Small Signals $\therefore$ Exclude static components $\rightarrow$ V_{ID} V_{IC} V_{DD} V_{SS} I_B

Superposition $\therefore$ Exclude other small signals $\rightarrow$ $v_{ic} \equiv v_{cc} \equiv v_{ee} \equiv i_b \equiv 0$

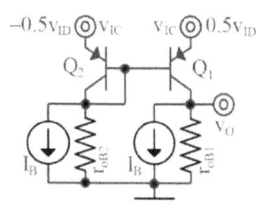 Small $\Rightarrow$ Signals 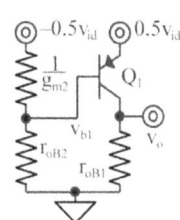 Approx. $\Rightarrow$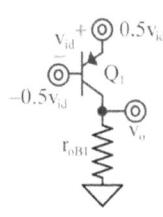

Parameters: $\quad R_{ID} \equiv \dfrac{v_{id}}{i_{id}} \approx \left(r_{o2} \,\|\, \dfrac{1}{g_{m2}} \,\|\, r_{\pi2} \right) + r_{\pi1} \approx r_{\pi1}$

$$G_D \equiv \left.\dfrac{i_o}{v_{id}}\right|_{v_o = 0} \approx g_{m1} \qquad\qquad v_{b1} = \dfrac{-0.5 v_{id}\left(r_{oB2} \,\|\, R_{B1}\right)}{1/g_{m2} + \left(r_{oB2} \,\|\, R_{B1}\right)} \approx -0.5 v_{id}$$

$$R_{OD1} \equiv \left.\dfrac{v_o}{i_1}\right|_{v_{id} = 0} = r_{o1} + R_S \qquad\qquad \pm v_{ID(MAX)} \equiv \pm \dfrac{\Delta i_{O(MAX)}}{G_D} \approx \pm \dfrac{I_B}{g_{m1}}$$

C. Small v_{IC}

Small Signals $\therefore$ Exclude static components → $V_{ID} \quad V_{IC} \quad V_{DD} \quad V_{SS} \quad I_B$

Superposition $\therefore$ Exclude other small signals → $v_{id} \equiv v_{cc} \equiv v_{ee} \equiv i_b \equiv 0$

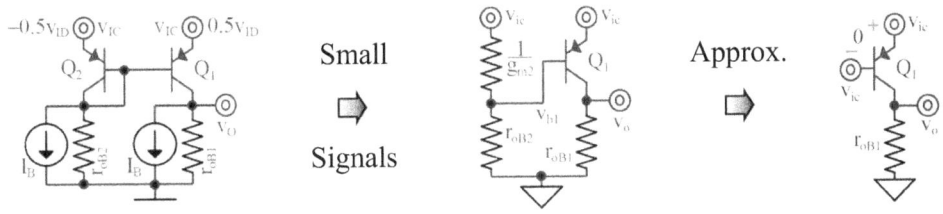

Parameters:
$$R_{IC1} \equiv \frac{v_{ic}}{i_{i1}} \approx r_{o1} + r_{oB1} \qquad v_{b1} = \frac{v_{ic}\left(r_{oB2} \| R_{B1}\right)}{1/g_{m2} + \left(r_{oB2} \| R_{B1}\right)} \approx v_{ic}$$

$$R_{IC2} \equiv \frac{v_{ic}}{i_{i2}} \approx \left(r_{o2} \| \frac{1}{g_{m2}} \| r_{\pi 2}\right) + r_{oB2} \approx r_{oB2}$$

$$G_C \equiv \left.\frac{i_o}{v_{ic}}\right|_{v_o = 0} \approx 0 \qquad R_{OC1} \equiv \left.\frac{v_o}{i_1}\right|_{v_{ic} = 0} = r_{o1}$$

2.5. Simulations: A. Electrical Components

Voltage Source: V[name] node+ node– dc=[value] ac=[value] [stimulus]

Current Source: I[name] source output dc=[value] ac=[value] [stimulus]

Resistor: R[name] node1 node2 [value]

Capacitor: C[name] node1 node2 [value]

Inductor: L[name] node1 node2 [value]

Diode: D[name] anode cathode [model]

BJT: Q[name] collector base emitter [model]

MOSFET: M[name] d g s b [model] W=[width] L=[length] M=[multiplier]

Voltage Amp: E[name] node+ node– vnode+ vnode– [gain]

Current Amp: F[name] source output [controlling voltage source] [gain]

Transconductor: G[name] source output vnode+ vnode– [gain]

Transimpedance: H[name] node+ node– [controlling voltage source] [gain]

B. Models

Nominal Diode: .model [name] d is=1f n=1 tt=100p cjo=100f vj=600m bv=7

Nominal BJT: .model [name] [npn or pnp] bf=100 va=50 is=1f
+ tf=100p cjc=100f vjc=600m mjc=0.5

Nominal MOS: .model [name] [nmos or pmos] vto=0.4 kp=200u lambda=10m
+ tox=5n ld=30n cgso=200p cgdo=200p ↳ or

* For almost no delay: tt = tf = 0 s & no cjo, tox, cgso, cgdo   "uo=290" [cm²/Vs]

C. Commands

ASCII Text File: [name].cir

First Line: [text]   Comment Lines: * [text]

Operating Point: .op

DC Sweep: .dc [source] [start] [end] [step]

Time-Domain Response: .tran [end]

Small-Signal Response: .ac dec [data points per decade] [start freq.] [end freq.]

Last Line (end of file):   .end

D. Example

$V_O \approx 2.5$ V

* PMOS Diff. Stage

$v_{IC} \geq 2.1$ V

vcc vcc 0 dc=5

≤ 3.9 V

qb vb vb vcc pnp1

rb vb 0 42.7k

$A_{D0} = 4.8$ V/V

qt vs vb vcc pnp1

$= 14$ dB

m1 0 vi1 vs vs pmos1 w=10u l=1u

m2 vo vi2 vs vs pmos1 w=10u l=1u

rd vo 0 50k

.model pnp1 pnp is=1f bf=100 va=75 rc=500

.model pmos1 pmos vto=-400m kp=40u lambda=20m

vidp vi1 vic dc=0 ac=0.5 sin 0 5m 100k

vidn vic vi2 dc=0 ac=0.5 sin 0 5m 100k

vic vic 0 dc=3 ac=0 sin 3 100m 1e6
                    ↑
* Higher noise?   Noise

.op $A_{C0} = -33$ mV/V

*.dc vic 0 5 10m $= -30$ dB

*.ac dec 100 100 1e6

*.tran 30u

.end

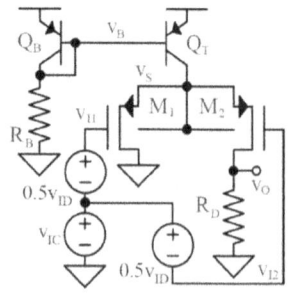

Chapter 3. Feedback

3.1. Feedback Loop

3.2. Impedances

3.3. Analysis

3.4. Configurations

3.5. Stability

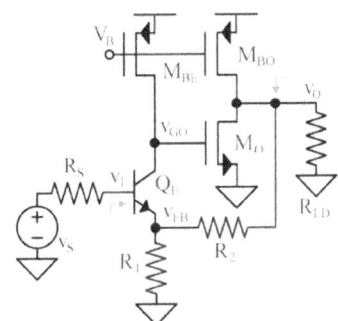

3.1. Feedback Loop: A. Model

Conceptual Model

Electrical Model

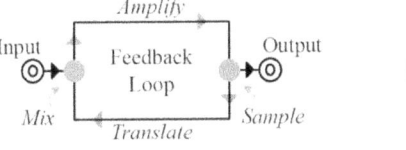

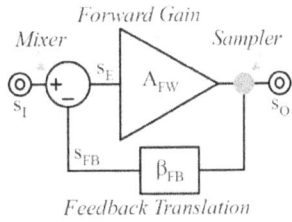

Mixer: s_I s_{FB} s_E $\rightarrow$ Same type $\therefore$ Same dimensional units

Gains: $A_{FW} \equiv \dfrac{s_O}{s_E}$ $\beta_{FB} \equiv \dfrac{s_{FB}}{s_O}$ + Loop Gain $\equiv A_{LG} = \dfrac{s_{FB}}{s_E} = A_{FW}\beta_{FB}$

B. Closed-Loop Translations

Error: $\quad s_E = s_I - s_{FB} = s_I - s_E A_{FW}\beta_{FB} = s_I - s_E A_{LG} = \dfrac{s_I}{1+A_{LG}}$

$\qquad$ If $\quad A_{LG} \to \infty \quad \therefore \quad s_E \to 0 \qquad\qquad \rightarrow \quad A_{LG}$ suppresses error

$\qquad\qquad\qquad\qquad\qquad\qquad\qquad\qquad\qquad\qquad$ Gain error $s_E \approx \dfrac{s_I}{A_{LG}}$

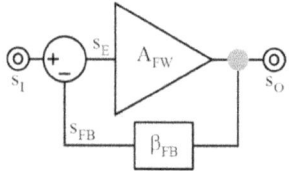

$\qquad\qquad\qquad\qquad\qquad\qquad\qquad s_{FB} \approx s_I \qquad \rightarrow \quad$ Mirrored reflection

$\qquad\qquad\qquad\qquad\qquad\qquad\qquad\qquad\qquad\qquad - $ FB opposes deviations

$\qquad\qquad\qquad\qquad\qquad\qquad\qquad s_O \approx \dfrac{s_I}{\beta_{FB}} \qquad \rightarrow \quad$ Feedback gain $\approx \dfrac{1}{\beta_{FB}}$

Gain: $\quad s_O = s_E A_{FW} = \left(s_I - s_{FB}\right)A_{FW} = \left(s_I - s_O\beta_{FB}\right)A_{FW} = \dfrac{s_I A_{FW}}{1+A_{FW}\beta_{FB}} = \dfrac{s_I A_{FW}}{1+A_{LG}}$

$$A_{CL} \equiv \dfrac{s_O}{s_I} = \dfrac{A_{FW}}{1+A_{FW}\beta_{FB}} = A_{FW} \,\|\, \dfrac{1}{\beta_{FB}} \approx \text{Lowest Input-to-Output Gain}$$

3.2. Impedances: A. Mixer

Series Mixer:

$\quad$ Voltages combine (add/subtract) in series $\qquad\qquad$ Dimension-less translation

$\quad \therefore \quad s_I \equiv v_I \qquad\qquad s_{FB} \equiv v_{FB}$

$$Z_{I CL} \equiv \dfrac{v_I}{i_{IN}} = \dfrac{v_E + v_{FB}}{i_{IN}} = \dfrac{v_E + v_E A_{LG}}{i_{IN}} = \left(\dfrac{v_E}{i_{IN}}\right)\left(1+A_{LG}\right) = Z_I + A_{LG}Z_I = Z_I + Z_{SER}$$

$\qquad\qquad\qquad\qquad\qquad\qquad\qquad\qquad\qquad\qquad\qquad\qquad\qquad$ Open-Loop Z_{IN} $(A_{LG} = 0)$

Shunt Mixer:

$\quad$ Currents combine in parallel

$\quad \therefore \quad s_I \equiv i_I \qquad\qquad s_{FB} \equiv i_{FB} \qquad Z_I$

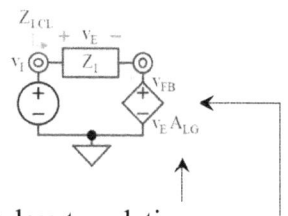

$\qquad\qquad\qquad\qquad\qquad\qquad\qquad\qquad\qquad\qquad\qquad\qquad Z_{SH} \equiv \dfrac{v_{IN}}{i_{FB}} = \dfrac{1}{G_{LGI}}$

$$Z_{I CL} \equiv \dfrac{v_{IN}}{i_I} = \dfrac{v_{IN}}{i_E + i_{FB}} = \dfrac{v_{IN}}{i_E + i_E A_{LG}} = \left(\dfrac{v_{IN}}{i_E}\right)\left(\dfrac{1}{1+A_{LG}}\right) = Z_I \,\|\, \dfrac{Z_I}{A_{LG}}$$

$\qquad\qquad\qquad\qquad\qquad\qquad\qquad\qquad\qquad\qquad\qquad\qquad\qquad\qquad$ Loop Gain

$\qquad\qquad\qquad\qquad\qquad\qquad\qquad\qquad\qquad\qquad\qquad\qquad\qquad\qquad$ Transconductance

B. Sampler

$A_{LG} \equiv$ Dimension-less translation

Series Sampler:

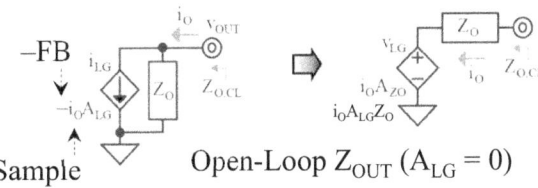

Ammeter senses in series

$\therefore \quad s_O \equiv i_O \quad \rightarrow \quad$ Norton i_O

Open-Loop Z_{OUT} ($A_{LG} = 0$)

$$Z_{O.CL} \equiv \frac{v_{OUT}}{i_O} = \frac{(i_O - i_{LG})Z_O}{i_O} = \frac{(i_O + i_O A_{LG})Z_O}{i_O} = Z_O + A_{LG}Z_O$$

$$\overbrace{}^{Z_{SER}}$$

Shunt Sampler:

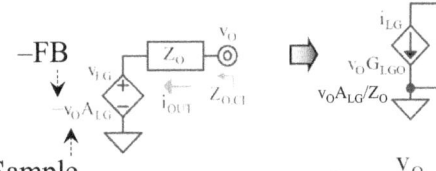

Voltmeter senses in parallel

$\therefore \quad s_O \equiv v_O \quad \rightarrow \quad$ Thévenin v_O

Sample

$$Z_{SH} \equiv \frac{v_O}{i_{LG}} = \frac{1}{G_{LGO}}$$

$$Z_{O.CL} \equiv \frac{v_O}{i_{OUT}} = \frac{v_Z + v_{LG}}{i_{OUT}} = \frac{i_{OUT}Z_O - v_O A_{LG}}{i_{OUT}} = \frac{Z_O}{1 + A_{LG}} = Z_O \parallel \frac{Z_O}{A_{LG}}$$

3.3. Analysis: A. Process

i. One Feedback Loop

Identification: s_I & s_{FB} from input terminal

s_O from output terminal

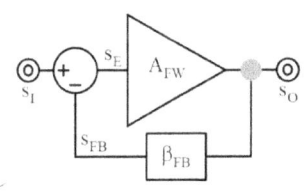

Feedback Model:

Decouple R's from i_I & i_{FB} $\rightarrow$ Extract Norton current sources for i_I & i_{FB}

Derive open-loop (two-port) parameters: R_I A_{FW} R_O β_{FB}

$\rightarrow$ Open loop: Keep all R's $\rightarrow$ Disable voltage/current sources

Keep inner loops closed

Forward Translation:

Calculate closed-loop parameters: $R_{I.CL}$ A_{CL} $R_{O.CL}$

Overall Response: Derive overall gain

B. Identification

Series: $s_{I/O}$ node $\neq$ In loop

v_I Mixers: Diff. Pairs   → $i_O = |v_I - v_{FB}|G_D$

Transistors   → $i_g = |v_i - v_{fb}|g_m$

i_O Samplers: Current Mirrors   → $i_O = A_I i_I$

Current Buffers   → $i_O \approx i_I$

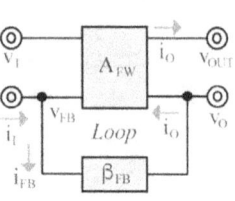

Shunt: $s_{I/O}$ node = In loop

"T" connections shunt-mix & shunt-sample

v_O Samplers: Diff. Pairs   → $i_O = |v_O - 0|G_D$ or $|0 - v_O|G_D$

Transistors   → $i_g = |v_o - 0|g_m$ or $|0 - v_o|g_m$

Source/Load $R_{S/LD}$:   Not part of series $R_{I/O}$ Part of shunt $R_{I/O}$

C. Signal Translation

Entire signal s_I to s_O if   Mixer senses s_E & β_{FB} translates s_O to s_{FB}

Small signal s_i to s_o if   Mixer senses s_e only   or β_{FB} translates s_o or s_{fb} only

3.4. Configurations: A. Transconductance: i. Degenerated CE/S

Identification

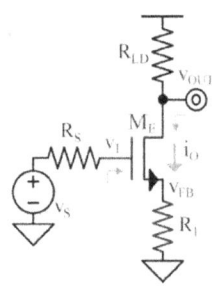

1. Inverting Feedback: Yes

If $i_O \uparrow$ $\therefore$ $v_{FB} \uparrow$ → $v_{GS} \downarrow$ → $i_O \downarrow$

2. Input Terminal: $v_I \neq$ In loop

3. M_E mixes v_i & v_{fb}

4. Output Terminal: $v_{OUT} \neq$ In loop   Loop = One node → v_{FB}

5. $v_{FB} = f(i_O)$ $\therefore$ M_E samples i_O $i_O \propto (v_I - v_{FB})^N$ → Nonlinear

6. Network: Series–Series $i_g \propto v_i - v_{fb}$ → Linear translation

7. Amplifier: $\dfrac{i_o}{v_i}$ → $A_{G.CL}$ $\therefore$ M_E mixes small signals only

Feedback Model:

$$R_I = R_{IN}\big|_{No\ g_m} \to \infty \qquad\qquad R_O \equiv R_{OUT}\big|_{No\ g_m} = r_{ds} + R_1$$

$$A_G \equiv \frac{i_o}{v_e}\bigg|_{v_{out}=0} = \frac{i_g + i_{rds}}{v_e} = \frac{v_e g_m + \left(-\dfrac{v_{fb}}{r_{ds}}\right)}{v_e} = \frac{v_e g_m - \dfrac{v_e g_m \left(R_1 \| r_{ds}\right)}{r_{ds}}}{v_e} = g_m\left(\frac{r_{ds}}{R_1 + r_{ds}}\right)$$

Where $\qquad\qquad\qquad M_E$ mixes v_e only $\quad\therefore\quad$ – FB translates v_i to i_o only

$$v_{fb}\big|_{v_{out}=0} = i_g \left(R_1 \| r_{ds}\right) = v_e g_m \left(R_1 \| r_{ds}\right)$$

$$\beta_{FB} \equiv \frac{v_{FB}}{i_O} = \frac{i_O R_1}{i_O} = R_1$$

Forward Translation:

$$A_{G.CL} = A_G \| \frac{1}{\beta_{FB}} = \frac{A_G}{1 + A_G \beta_{FB}} = \frac{g_m\left(\dfrac{r_{ds}}{R_1 + r_{ds}}\right)}{1 + g_m\left(\dfrac{r_{ds}R_1}{R_1 + r_{ds}}\right)} = \frac{g_m}{\left(\dfrac{R_1 + r_{ds}}{r_{ds}}\right) + g_m R_1}$$

$$= \frac{g_m}{\dfrac{R_1}{r_{ds}} + 1 + g_m R_1} = \frac{g_m}{1 + \left(g_m + \dfrac{1}{r_{ds}}\right) R_1}$$

$$R_1 \to \infty \qquad \therefore \qquad R_{I.CL} \to \infty$$

$$R_{O.CL} = R_O \left(1 + A_G \beta_{FB}\right) = \left(r_{ds} + R_1\right)\left[1 + g_m\left(\frac{r_{ds}R_1}{R_1 + r_{ds}}\right)\right] = r_{ds} + R_1 + g_m r_{ds} R_1$$

$\rightarrow$ Results match those from direct analysis.

ii. v_I-Mixed Current Source: Regulated-Cascode Transconductor

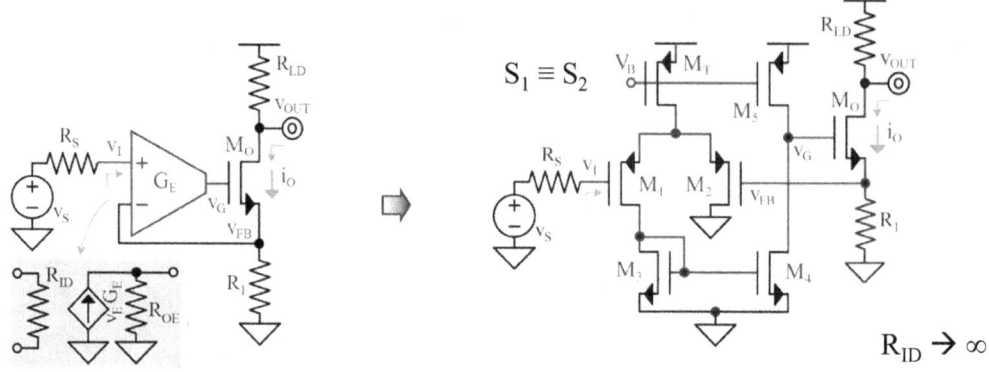

$$S_1 \equiv S_2$$

$$R_{ID} \to \infty$$

Outer Loop: Gate-Looped M_O $G_E \equiv \dfrac{i_4}{v_{ID}}\bigg|_{v_g=0} \approx \left(\dfrac{-g_{m1}}{2}\right)\left(\dfrac{1}{g_{m3}}\right)(-g_{m4}) = \dfrac{g_{m1}g_{m4}}{2g_{m3}}$

Entwined Loop: Source-Degenerated M_O

Note: High $R_{O.CL}$ $\therefore$ Current Source $R_{OE} \equiv \dfrac{v_g}{i_g}\bigg|_{v_{ID}=0} = r_{ds4} \,\|\, r_{ds5}$

Identification

1. Inverting Feedback: Yes   4. Output Terminal: $v_{OUT} \neq$ In loop

   If $v_{FB} \uparrow$ $\therefore$ $v_G \downarrow \to i_O \downarrow \to v_{FB} \downarrow$ 5. $v_{FB} = f(i_O)$ $\therefore$ M_O samples i_O

2. Input Terminal: $v_I \neq$ In loop   6. Network: Series–Series

3. G_E mixes v_I & v_{FB} 7. Amplifier: $\dfrac{i_O}{v_I}$ $\to A_{G.CL}$

Feedback Model:

$$R_I = R_{IN}\big|_{No\ G_E} = R_{ID} + \left(R_1 \,\|\, R_{SO}\right) \qquad\qquad R_O = R_{OUT}\big|_{No\ G_E} = R_{DO}$$

$$A_G \equiv \dfrac{i_o}{v_E}\bigg|_{v_{out}=0} = G_E R_{OE} G_O \qquad \beta_{FB} \equiv \dfrac{v_{FB}}{i_o} = R_1 \,\|\, \left(R_{ID} + R_S\right)$$

G_E senses v_E

β_{FB} translates i_O to v_{FB} } – FB translates v_I to i_O

If $R_{ID} + R_S \gg R_1$

Loaded Gain: $\dfrac{i_{LD}}{v_S} = \left(\dfrac{R_{ICL}}{R_S + R_{ICL}}\right)\left(A_G \,\|\, \dfrac{1}{\beta_{FB}}\right)\left(\dfrac{-R_{O.CL} \,\|\, R_{LD}}{-R_{LD}}\right) \approx \dfrac{1}{\beta_{FB}} \approx \dfrac{1}{R_1}$

34

B. Voltage: i. Non-Inverting Amp

Outer Loop:

    Gate-Looped M_O

Entwined Loop:

    Emitter-Degenerated Q_E

Identification:

Q_E:M_O Amp:

$$R_{ID} = r_{\pi E}$$

$$G_E = -g_{mE}R_{GO}(-g_{mO})$$

$$R_{EO} = r_{dsO} \,\|\, r_{dsBO}$$

1. Inverting Feedback: Yes

    If $v_{FB} \uparrow$ $\therefore$ $v_{GO} \uparrow \rightarrow v_O \downarrow \rightarrow v_{FB} \downarrow$

2. Input Terminal: $v_I \neq$ In loop

3. Q_E mixes v_i & v_{fb}

4. Output Terminal: $v_O =$ In loop

5. $v_{FB} = f(v_O)$ $\therefore$ R_2 T-samples v_O

6. Network: Series–Shunt

7. Amplifier: $\dfrac{v_o}{v_i} \rightarrow A_{V.CL}$

Note: Q_E:M_O amplifies v_e like non-inverting op amp $\rightarrow$ Equivalent analysis

Feedback Model: $\quad R_I = R_{IN}\big|_{No\ g_{mO}} = R_{BE} \qquad R_O = R_{OUT}\big|_{No\ g_{mO}} = R_{OE} \,\|\, R_2'' \,\|\, R_{LD}$

Where: $\qquad\qquad R_2' = R_2 + (R_{OE} \,\|\, R_{LD}) \qquad\qquad R_2'' = R_2 + (R_1 \,\|\, R_{EE})$

$$A_V \equiv \frac{v_o}{v_e} \approx -g_{mE}(r_{dsBE} \,\|\, R_{CE})(-g_{mO})R_O + g_{mE}(R_1 \,\|\, R_2' \,\|\, R_{EE})\left(\frac{R_{OE} \,\|\, R_{LD}}{R_2'}\right)$$

$$\uparrow$$
$$R_2 \text{ feeds } i_{gE}{:}i_2 \text{ to } v_O$$

$$\approx g_{mE}r_{dsBE}g_{mO}R_O$$

$$\beta_{FB} \equiv \frac{v_{fb}}{v_o} = \frac{R_1 \,\|\, R_{EE}}{R_2 + (R_1 \,\|\, R_{EE})} \approx \frac{R_1}{R_1 + R_2} \quad \text{If } R_1 \ll R_{EE}$$

Q_E senses v_e only $\therefore$ $-$ FB translates v_i to v_o only

Loaded Gain: $\qquad$ Non-Inv. Op-Amp Gain

$$\frac{v_o}{v_s} = \left(\frac{R_{ICL}}{R_S + R_{ICL}}\right)\left(A_V \,\|\, \frac{1}{\beta_{FB}}\right) \approx \frac{1}{\beta_{FB}} \approx \frac{R_1 + R_2}{R_1}$$

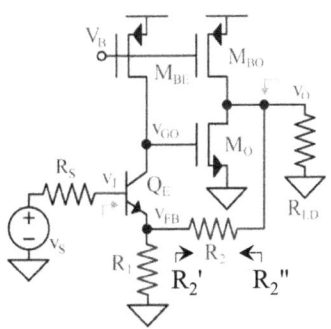

ii. Gate-Coupled Amp

Outer Loop: Gate-Looped M_6

Entwined Loop: Source-Degenerated M_2

Inner Loop: Source-Degenerated M_4

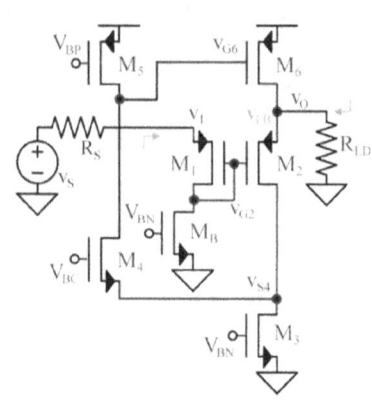

Identification

1. Inverting Feedback: Yes

    If $v_O \uparrow$ $\therefore$ $v_{S4} \uparrow \rightarrow v_{G6} \uparrow \rightarrow v_O \downarrow$

2. Input Terminal: $v_I \neq$ In loop

3. M_{12} mixes v_I & v_{FB}

4. Output Terminal: v_O = In loop

5. $v_{FB} = f(v_O)$ $\therefore$ M_{12} T-samples v_O

6. Network: Series–Shunt

7. Amplifier: $\dfrac{v_O}{v_I} \rightarrow A_{V.CL}$

Feedback Model:

$$R_I = R_{ID}\big|_{No\ g_{m6}} \rightarrow \infty \qquad R_{IN} = \frac{1}{g_{m1}} + \left(r_{dsB} \parallel R_{G2}\right) \approx r_{dsB}$$

$$R_O = R_{OUT}\big|_{No\ g_{m6}} = r_{ds6} \parallel R_{S2} \parallel R_{LD} \approx R_{S2} \parallel R_{LD}$$

M_{12} feeds i_{g12} to v_O
$\downarrow$

$$A_V \equiv \frac{v_o}{v_E} = -g_{m2}\left(r_{ds3} \parallel R_{S4} \parallel R_{D2}\right)\left(\frac{1}{R_{S4}}\right)r_{ds5}(-g_{m6})R_O + g_{m2}R_O$$

$$\approx g_{m2}r_{ds5}g_{m6}R_O$$

$$\beta_{FB} \equiv \frac{v_{FB}}{v_O} = 1 \qquad R_{SHO} = \frac{1}{G_{LGO}} \approx \frac{1}{g_{m2}r_{ds5}g_{m6}} \qquad G_{LGO} = \frac{i_{g6}}{v_o} \approx \left(\frac{1}{R_{S2}}\right)r_{ds5}g_{m6}$$

M_{12} senses v_E & β_{FB} translates v_O to v_{FB} $\quad\therefore\quad$ $-$FB translates v_I to v_O

Loaded Gain: $\dfrac{v_O}{v_S} = \left[\dfrac{R_{IN} \parallel R_{LCL}}{R_S + \left(R_{IN} \parallel R_{LCL}\right)}\right]\left(A_V \parallel \dfrac{1}{\beta_{FB}}\right) \approx \dfrac{1}{\beta_{FB}} = 1 \qquad$ If $R_S \ll R_{IN}$

C. Transimpedance: i. Diode-Connected Transistor

Identification

1. Inverting Feedback: Yes

   If $v_O \uparrow$ ∴ $i_C \uparrow$ → $v_O \downarrow$

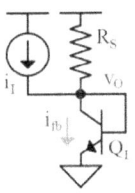

2. Input Terminal: v_O = In loop

3. v_O T-mixes i_I & i_C

4. Output Terminal: v_O = In loop   Loop = One node → v_O

5. $i_{fb} = f(v_o)$ ∴ Q_I T-samples v_o $i_C \propto \exp v_O$

6. Network: Shunt–Shunt $i_g = i_{fb} \propto v_o$ → Linear translation

7. Amplifier: $\dfrac{v_o}{i_I}$ → $A_{Z.CL}$ ∴ Q_I T-samples small signals only

Feedback Model:

$$R_I = R_{IN}\big|_{No\ g_m} = r_\pi \| r_o \| R_S \qquad R_O = R_{OUT}\big|_{No\ g_m} = r_\pi \| r_o \| R_S \qquad \beta_{FB} \equiv \frac{i_{fb}}{v_o} = g_m$$

$$A_Z \equiv \frac{v_o}{i_e} = r_\pi \| r_o \| R_S$$

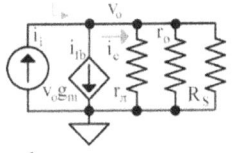

Forward Translation:

$$R_{SHI/O} = \frac{1}{G_{LGI/O}} = \frac{v_o}{i_{fb} + i_e} = \frac{1}{g_m} \| r_\pi \| r_o \| R_S$$

$$R_I = R_O = A_Z \qquad ∴ \qquad R_{I.CL} = R_{O.CL} = \frac{R_{I/O}}{1 + A_Z \beta_{FB}} = A_{Z.CL} = A_Z \| \frac{1}{\beta_{FB}} = r_\pi \| r_o \| R_S \| \frac{1}{g_m}$$

→ Results match those from direct analysis.

v_O T-senses i_E & Q_I samples v_o only ∴ – FB translates i_i to v_o only

ii. Inverting Amp

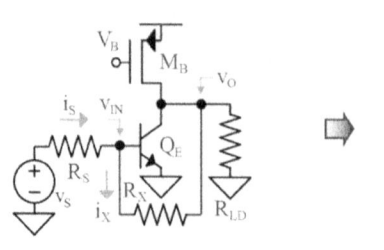 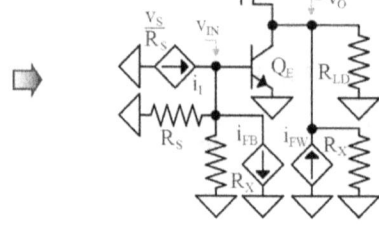

Identification

1. Inverting Feedback: Yes

    If $v_{IN} \uparrow$ ∴ $v_O \downarrow$ → $v_{IN} \downarrow$

2. Input Terminal: v_{IN} = In loop

3. v_{IN} T-mixes i_I & i_{FB}

4. Output Terminal: v_O = In loop

5. $i_{FB} = f(v_O)$ ∴ R_X T-samples v_O

6. Network: Shunt–Shunt

7. Amplifier: $\dfrac{v_O}{i_I}$ → $A_{Z.CL}$

Inv. Q_E Amp: $R_{ID} = r_{\pi E}$ $G_E = g_{mE}$ $R_{EO} = r_{oE} \parallel r_{dsB}$

Note: Q_E amplifies v_s like inverting op amp → Equivalent analysis

Feedback Model: $R_I = R_{IN}\big|_{No\ i_{FB}} = R_S \parallel R_X \parallel R_{ID}$ $R_O = R_{OUT}\big|_{No\ i_{FB}} = R_{EO} \parallel R_X \parallel R_{LD}$

$$A_Z \equiv \frac{v_o}{i_e} = R_I\left[(-g_{mE}) + \left(\frac{1}{R_X}\right)\right]R_O \qquad \beta_{FB} \equiv \frac{i_{FB}}{v_O} = -\frac{1}{R_X}$$

     $\rightarrow R_X$ feeds i_I:i_{FW} to v_O

$\approx -R_I g_{mE} R_O$

v_{IN} T-senses i_E

β_{FB} translates v_O to i_{FB}

$$R_{SHO} = \frac{1}{G_{LGO}} = \frac{v_o}{i_{gE}} = \frac{1}{(1/R_X)R_I g_{mE}}$$

→ – FB translates i_I to v_O

Loaded A_V Gain: $\dfrac{v_o}{v_s} = \left(\dfrac{1}{R_S}\right)\left(A_Z \parallel \dfrac{1}{\beta_{FB}}\right) \approx -\dfrac{R_X}{R_S}$

Inv.
Op-Amp
Gain

→   Results match those from direct analysis.   Bias Note: $V_s \approx V_{BE} = V_O\big|_{i_R\text{'s}\approx 0}$

D. Current: i. i_I-Mixed Regulated-Cascode Current Mirror

Series Loop:

    Diode-Connected Q_1

Outer Loop: Gate-Looped M_O

Entwined Loop:

    Source-Degenerated M_O

Identification

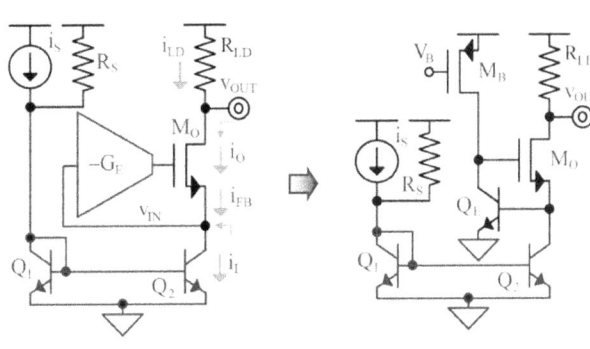

1. Inverting Feedback: Yes

    If $v_{IN} \uparrow$ $\therefore$ $v_{GO} \downarrow$ $\rightarrow$ $v_{IN} \downarrow$

2. Input Terminal: v_{IN} = In loop

3. v_{IN} T-mixes i_I & i_{FB}

4. Output Terminal: $v_{OUT} \neq$ In loop

5. $i_{FB} = f(i_O)$ $\therefore$ M_O samples i_O

6. Network: Shunt–Series

7. Amplifier: $\dfrac{i_O}{i_I}$ $\rightarrow$ $A_{I.CL}$

Note: M_O's $v_{gsO}g_{mO}$ does not feed outer loop $\therefore$ $M_O \neq$ Voltage Mixer

Feedback Model:

$$R_I = R_{IN}\big|_{No\ G_F} = r_{o2}\ \|\ R_{EI}\ \|\ R_{SO} \approx R_{SO} \qquad R_O = R_{OUT}\big|_{No\ G_F} = R_{DO} \qquad \beta_{FB} \equiv \frac{i_{FB}}{i_O} = 1$$

$$A_I \equiv \frac{i_o}{i_c}\bigg|_{v_{out}=0} = \frac{-(r_{o2}\ \|\ R_{EI})(-G_E)R_{EO}\,g_{mO}}{1 + \left(g_{mO} + \dfrac{1}{r_{dsO}}\right)(r_{o2}\ \|\ R_{EI})} \approx G_E R_{EO}$$

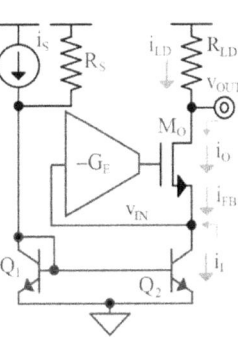

$$R_{SHI} = \frac{1}{G_{LGI}} \approx \frac{1}{G_E R_{EO} G_O} \qquad G_{LGI} = \frac{-i_{fb}}{v_{in}} \approx -G_E R_{EO}(-G_O)$$

v_{IN} T-senses i_E & β_{FB} translates i_O to i_{FB} $\therefore$ $-$ FB translates i_I to i_O

$$\text{If }\ \frac{1}{g_{m1}} << R_S$$

$$\downarrow$$

Loaded Gain: $\dfrac{i_{LD}}{i_s} = \left(\dfrac{i_I}{i_s}\right)\left(\dfrac{i_O}{i_I}\right)\left(\dfrac{i_{LD}}{i_O}\right) \approx \left(\dfrac{A_{E2}}{A_{E1}}\right)\left(A_I\ \|\ \dfrac{1}{\beta_{FB}}\right)\left[\dfrac{-(R_{O.CL}\ \|\ R_{LD})}{-R_{LD}}\right] \approx \dfrac{A_{E2}}{A_{E1}}$

ii. Current Amp

Outer Loop: Gate-Looped M_O

Entwined Loops:

    Source-Degenerated M_3

    Source-Degenerated M_O

Inner Loop: Diode-Connected M_5

Identification

1. Inverting Feedback: Yes

    If $v_{GO} \uparrow$ $\therefore$ $v_{G34} \uparrow \rightarrow v_{IN} \downarrow \rightarrow v_{GO} \downarrow$

2. Input Terminal: v_{IN} = In loop

3. v_{IN} T-mixes i_I & i_{FB}

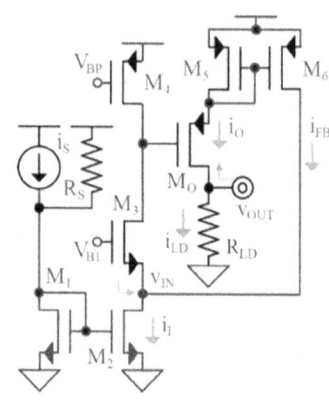

4. Output Terminal: $v_{OUT} \neq$ In loop

5. $i_{FB} = f(i_O)$ $\therefore$ M_O samples i_O

6. Network: Shunt–Series

7. Amplifier: $\dfrac{i_O}{i_I} \rightarrow A_{I.CL}$

Bias: I_4 offsets i_I $\rightarrow$ $\quad i_O \approx \dfrac{i_I}{\beta_{FB}} = \dfrac{i_2 - I_4}{\beta_{FB}} \approx \left[i_s \left(\dfrac{S_2}{S_1} \right) - I_4 \right] \left(\dfrac{1}{\beta_{FB}} \right)$ $\quad$ If R_S is high

Feedback Model:

$$R_I = R_{IN}\big|_{No\ g_{m6}} = r_{ds2} \| R_{S3} \| r_{ds6} \approx R_{S3} \qquad\qquad R_O = R_{OUT}\big|_{No\ g_{m6}} = R_{DO}$$

$$A_I \equiv \dfrac{i_o}{i_e}\bigg|_{v_{out}=0} = (-R_I)\left(\dfrac{1}{R_{S3}}\right) r_{ds4}(-G_O) \approx r_{ds4}G_O \qquad\qquad \beta_{FB} \equiv \dfrac{i_{FB}}{i_O} \approx \dfrac{(W/L)_6}{(W/L)_5} \equiv \dfrac{S_6}{S_5}$$

$$R_{SHI} = \dfrac{1}{G_{LGI}} \approx \dfrac{R_{S3}}{r_{ds4}G_O S_6/S_5} \qquad G_{LGI} = \dfrac{-i_{fb}}{v_{in}} \approx \left(\dfrac{1}{R_{S3}}\right) r_{ds4}(-G_O)\left(-\dfrac{S_6}{S_5}\right)$$

v_{IN} T-senses i_E

                    } – FB translates i_I to i_O

β_{FB} translates i_O to i_{FB}

Loaded Gain: If $\dfrac{1}{g_{m1}} \ll R_S$

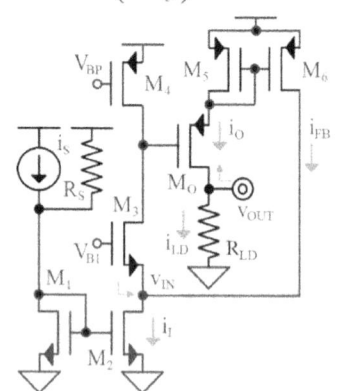

$$\dfrac{i_{ld}}{i_s} \approx \left(\dfrac{S_2}{S_1}\right)\left(A_I \| \dfrac{1}{\beta_{FB}}\right)\left(\dfrac{R_{O.CL} \| R_{LD}}{R_{LD}}\right) \approx \left(\dfrac{S_2}{S_1}\right)\left(\dfrac{S_5}{S_6}\right)$$

E. Positive Feedback

Shunt – FB: $Z_{SH} = +\dfrac{1}{G_{LG}}$ → i_Z opposes rise in v_Z

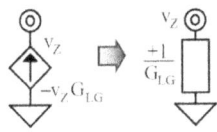

Shunt + FB: $Z_{SH} = -\dfrac{1}{G_{LG}}$ → i_Z reinforces rise in v_Z

$\therefore$ + FB feeds/supplies parallel Z_X

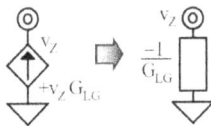

Circuit: $A_{LG+} = A_{FW}\beta_{FB} = G_{LG}Z_X = 1$ → Oscillates

$A_{LG+} > 1$ → Latches

$A_{LG+} < 1$ → Removes Z_X → $Z_{CL} = Z_X \parallel Z_{SH} = Z_X \parallel \dfrac{-1}{G_{LG}}$

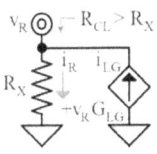

Applications:

Increase R_O → Increase gain $A_{V/Z}$ $= \left.\dfrac{Z_X}{1-G_{LG}Z_X}\right|_{A_{LG-}<1} > Z_X$

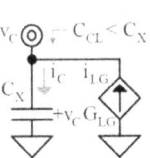

Reduce C_O → Extend bandwidth p_{BW}

Resupply (cancel) R_L loss → Sustain LC oscillations

i. Increase Resistance: CB/G Current Buffer

FB Loop: r_{ds} closes FB loop to v_S

Identification

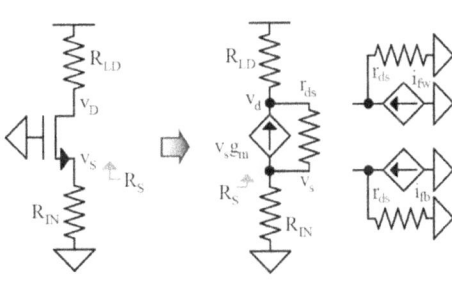

+ FB: $v_S \uparrow \to i_g \uparrow \to v_D \uparrow \to i_{rds} \uparrow \to v_S \uparrow$

v_S = In loop

v_S T-mixes i_i & T-samples v_s

Split r_{ds} with 2-port N:N model:

r_{ds} at v_s $\qquad G_{FB} = \dfrac{1}{r_{ds}}$ to v_s $\qquad r_{ds}$ at v_d $\qquad G_{FW} = \dfrac{1}{r_{ds}}$ to v_d

Closed-Loop Impedance:

$$R_S = R_{gm} \parallel r_{ds} \parallel R_{SH} = \dfrac{1}{g_m} \parallel r_{ds} \parallel \dfrac{1}{G_{LGS}} = \dfrac{r_{ds}}{1+g_m r_{ds}} \parallel \left(\dfrac{r_{ds}}{g_m r_{ds}+1}\right)\left(\dfrac{r_{ds}+R_{LD}}{-R_{LD}}\right) = \dfrac{r_{ds}+R_{LD}}{1+g_m r_{ds}}$$

$$G_{LGS} = \dfrac{-i_{fb}}{v_s} = \left(g_m + G_{FW}\right)\left(R_{LD} \parallel r_{ds}\right)\left(-G_{FB}\right) = \left(g_m + \dfrac{1}{r_{ds}}\right)\left(R_{LD} \parallel r_{ds}\right)\left(\dfrac{-1}{r_{ds}}\right)$$ Loaded R_{SG}

Cross-Coupled Pair

FB Loop: $M_5:M_6$ closes FB loop to v_{O1} & v_{O2}

Identification

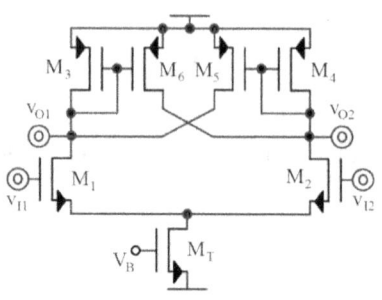

+ FB: $v_{O1} \uparrow \rightarrow v_{O2} \downarrow \rightarrow v_{O1} \uparrow$

v_{O1} & v_{O2} = In loop

v_{O12} T-mixes i_{12} & T-samples v_{o12}

Feature: Higher R_{O12} with shorter L_{CH} $\therefore$ Higher GBW = $A_V p_{BW}$

Gain: $\dfrac{v_{O12}}{v_{ID}} \approx \left(\dfrac{g_{m12}}{2}\right)R_{O12} > \left(\dfrac{g_{m12}}{2}\right)\left(\dfrac{1}{g_{m34}}\right)$

$R_{O1} \approx \dfrac{1}{g_{m3}} \parallel \dfrac{1}{G_{LGO1}} = \dfrac{1/g_{m3}}{1+\left(G_{LGO1}/g_{m3}\right)}$ $\qquad$ $G_{LGO1} = \dfrac{i_{g5}}{v_{o1}} \approx -g_{m6}\left(\dfrac{1}{g_{m4}}\right)g_{m5}$

$\qquad = \dfrac{1/g_{m3}}{1+A_{LG}} = \dfrac{1/g_{m3}}{1-\left(g_{m6}/g_{m4}\right)\left(g_{m5}/g_{m3}\right)} = \dfrac{1/g_{m3}}{1-\left(S_5/S_4\right)\left(S_6/S_3\right)} > \dfrac{1}{g_{m3}}$ If $A_{LG} < 1$

ii. Reduce Capacitance: CC/D Voltage Follower

FB Loop: C_{GS} closes FB loop to v_G & v_S

Identification

+ FB: $v_G \uparrow \rightarrow v_S \uparrow \rightarrow i_{fb} \uparrow \rightarrow v_G \uparrow$

v_G & v_S = In loop $\rightarrow$ v_G T-mixes i_i & T-samples v_g

Split C_{GS} with 2-port N:N model:

$\qquad$ C_{GS} at v_G $\qquad$ $G_{FW} = sC_{GS}$ to v_S $\qquad$ C_{GS} at v_S $\qquad$ $G_{FB} = sC_{GS}$ to v_G

Closed-Loop Impedance:

$Z_G = Z_{IN} \parallel Z_{GS} \parallel \dfrac{1}{G_{LGG}} \approx \left(R_{IN} \parallel \dfrac{1}{sC_{IN}}\right) \parallel \dfrac{1}{sC_{GS}} \parallel \dfrac{-1}{sC_{GS}} = R_{IN} \parallel \dfrac{1}{sC_{IN}}$ $\rightarrow$ C_{GS} cancels

$G_{LGG} = \dfrac{i_c}{v_g} = \left\{\dfrac{g_m}{1+\left[g_m+\left(1/r_{ds}\right)\right]\left(Z_{LD} \parallel Z_{GS}\right)} + G_{FW}\right\}\left(Z_{LD} \parallel Z_{GS}\right)\left(-G_{FB}\right) \approx -G_{FB} = -sC_{GS}$

Insight: $v_g \approx v_s$ $\qquad \therefore \qquad$ $i_c = v_{gs}sC_{GS} \approx 0$ $\qquad \rightarrow \qquad$ No C_{GS} in p_G or p_S

$\qquad$ $C_{G/S}$ shunts $v_{g/s}$ past $p_{G/S}$ $\qquad \rightarrow \qquad$ $i_{fb/fw}$ fades $\qquad \rightarrow \qquad$ $-sC_{GS}$ fades past $p_{G/S}$

Looped Follower

FB Loop: M_1:C_C:M_2 closes FB loop to v_{IN}

Identification

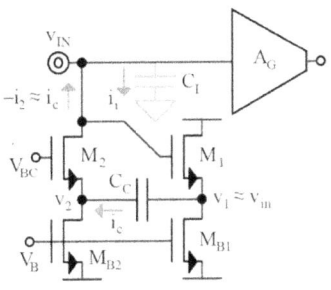

+ FB: $\quad v_{IN} \uparrow \rightarrow v_{SFB} \uparrow \rightarrow i_c \uparrow \rightarrow v_{IN} \uparrow$

$\qquad\quad v_{IN}$ = In loop

$\qquad\quad v_{IN}$ T-mixes i_i & T-samples v_{in}

Closed-Loop Impedance: R_{D2} blocks v_{in}:v_2:i_{fw} forward path

$$Z_{IN} = Z_1 \parallel R_{D2} \parallel \frac{1}{G_{LG}} \approx \frac{1}{sC_I} \parallel R_{D2} \parallel \frac{-1}{sC_C} = R_{D2} \parallel \frac{1}{s(C_1 - C_C)} \quad \rightarrow \quad C_C \text{ reduces } C_{IN}$$

$$G_{LG} = \frac{i_2}{v_{in}} = \left(\frac{v_1}{v_{in}}\right)\left(\frac{i_c}{v_1}\right)\left(\frac{i_2}{i_c}\right) \approx \frac{-1}{Z_C + (1/g_{m2})} \approx -sC_C \begin{bmatrix} \text{if } G_1\text{'s } g_{m1}Z_C \gg 1 \\[2mm] Z_C \gg 1/g_{m2} \end{bmatrix} \text{Up to } \frac{g_{m12}}{2\pi C_C}$$

3.5. Stability: A. Stability Criterion

Gain Objective: $\qquad A_{CL} \approx \dfrac{1}{\beta_{FB}} \geq 1 \qquad \therefore \qquad A_{FW} \gg \dfrac{1}{\beta_{FB}} \qquad \beta_{FB} \leq 1$

Stability Criterion: If p_1 p_2 < $f_{180°} \equiv$ Inversion f_O < $f_{0dB} \equiv$ Unity-Gain f_O

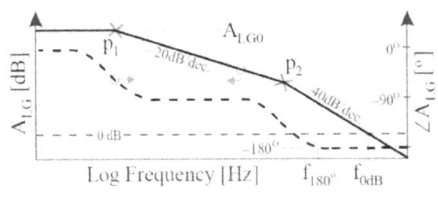 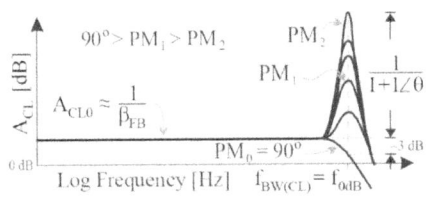

$A_{LG} = A_{FW}\beta_{FB} \qquad\qquad$ At $f_{0dB} \qquad$ When $\theta = 180° \qquad\qquad\qquad f_{BW(CL)} = f_{0dB}$

$$\therefore A_{CL} = A_{FW} \parallel \frac{1}{\beta_{FB}} = \frac{A_{FW}}{1 + A_{LG}} = \frac{1\angle\theta/\beta_{FB}}{1 + 1\angle\theta} = \frac{-1/\beta_{FB}}{1-1} \rightarrow \infty \rightarrow \text{Uncontrolled} \quad f_{0dB} < f_{180°}$$

Loop inverts at $f_{180°}$ $\rightarrow$ + FB $\qquad$ Phase (inversion) Margin $\equiv 180° + \angle A_{LG(0dB)} > 0°$

"Stable" if phase shift < 180° $\qquad$ Gain (suppression) Margin $\equiv 0$ dB $- A_{LG(180°)} > 0$ dB

B. Stabilized Response

If: $p_1 = \text{Low}$ $p_2 \approx f_{0dB}$ $p_{234\cdots N} \gg f_{0dB}$ $\therefore$ $\text{PM} \approx 45°$

Below f_{0dB}:   In-phase z_X's recover phase lost to intermediate p_X's

Power-Up: $A_{LG} = 0$ when $v_{DD} = 0$ $\rightarrow$ $f_{0dB} \propto A_{LG} \propto v_{DD}$ $\therefore$ f_{0dB} crosses f_O

$\therefore$ Ensure $\angle A_{LG} > -180°$ $\rightarrow$ Okay if $p_X < z_X$, but within 1 decade

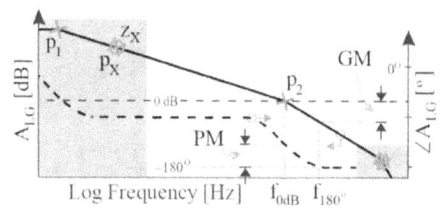   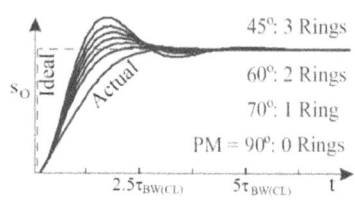

Ideal Step Response: s_O follows $\dfrac{s_I}{\beta_{FB}}$ $\rightarrow$ No delay   No ringing

Actual Response: p_1 delays s_O & p_2 reinforces s_O (w/ + FB)

Unacceptable Response: Excessive ringing $\rightarrow$ E.g.: PM < 20°   "Rings"

C. Stabilization

Goal: Reach f_{0dB} with $\angle A_{LG} > -180°$ High C_1 Low $\dfrac{1}{G_{LG2}}$

Strategies:

1.   Decrease p_1 $\rightarrow$ Add C_S to v_1

2.   Split $p_1 : p_2$ $\rightarrow$ Add $-$ FB C_S between v_1 & v_2

3.   Increase p_2 $\rightarrow$ Loop + FB C_S to v_2 $\rightarrow$ Low C_2

4.   Recover p_2's phase $\rightarrow$ Add in-phase zero

5.   Out-of-phase zeros:

   Shift to high f_O

   Convert to in-phase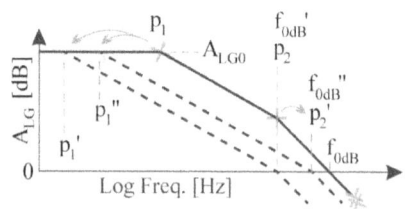

   Eliminate

44

Loop-Gain Simulation

Open loop: $\qquad\qquad\qquad\qquad\qquad\qquad\qquad\qquad$ $v_{XO} : v_{XI}$

Bias (reconnect without closing small-signal loop): $\qquad$ High L_{DC}

Load (reconstruct without altering bias): $\qquad\qquad$ Z_{IN}' $\qquad$ High C_{ACO}

Input (inject small signal into v_X without altering bias): v_{ac} $\qquad$ High C_{ACI}

Simulate: $\quad$ E.g., $\quad L_{DC} = 1$ MH $\qquad C_{AC}$'s $= 1$ MF $\qquad A_{LG} = \dfrac{v_{xo}}{v_{xi}}$ $\qquad \angle A_{LG0} = \pm 180°$

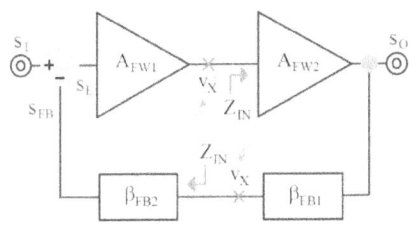 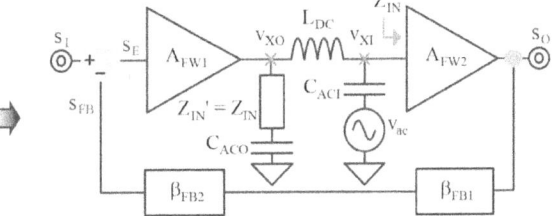

Convenient Break Points: $\quad$ Gates $\rightarrow$ $\quad R_{IN} \rightarrow \infty$ $\quad \therefore \quad Z_{IN} = \dfrac{1}{sC_G}$ $\quad$ No C_{ACO}

Verify (small- & large-signal) Stability:

Disturb closed loop $\quad \rightarrow \quad$ Apply sudden & wide $\quad \Delta s_I \qquad \Delta i_{LOAD} \qquad \Delta v_{DD/SS}$

$\qquad\qquad\qquad\qquad\qquad$ Stable if s_O recovers & settles after acceptable delay

D. Out-of-Phase Zero

Shift to High f_O:

Impede/reduce i_C $\quad \rightarrow \quad$ Nulling R_F impedes i_C

$\therefore \quad i_C > i_A$ at higher f_O $\quad \rightarrow \quad$ Higher z_F

$\qquad$ Short-circuit A_G incorporates z_F (when $v_O = 0$):

$$i_C = \left.\frac{v_{IN} - v_O}{1/sC_F + R_F}\right|_{v_O=0} \ge i_A = v_{IN}A_G \qquad \rightarrow \qquad f_O \ge \frac{1}{2\pi C_F\left(\dfrac{1}{A_G} - R_F\right)} = z_F$$

Convert to In-Phase: $\qquad\qquad$ R_F reduces effective R $\leftarrow$

If $\quad R_F \ge \dfrac{1}{A_G}$ $\quad \therefore \quad i_C$ cannot surpass i_A $\quad \rightarrow \quad$ No out-of-phase zero $\quad \rightarrow \quad$ Shift to ∞

R_F current-limits C_F $\quad \therefore \quad C_F$ effect on p_{IN} fades $\quad \rightarrow \quad$ Adds reversal in-phase zero

Eliminate:

Buffering – FB signal   →   Increases C_{IN}, Reduces Z_O

Shunting or blocking FW i_C →   Eliminates z_F

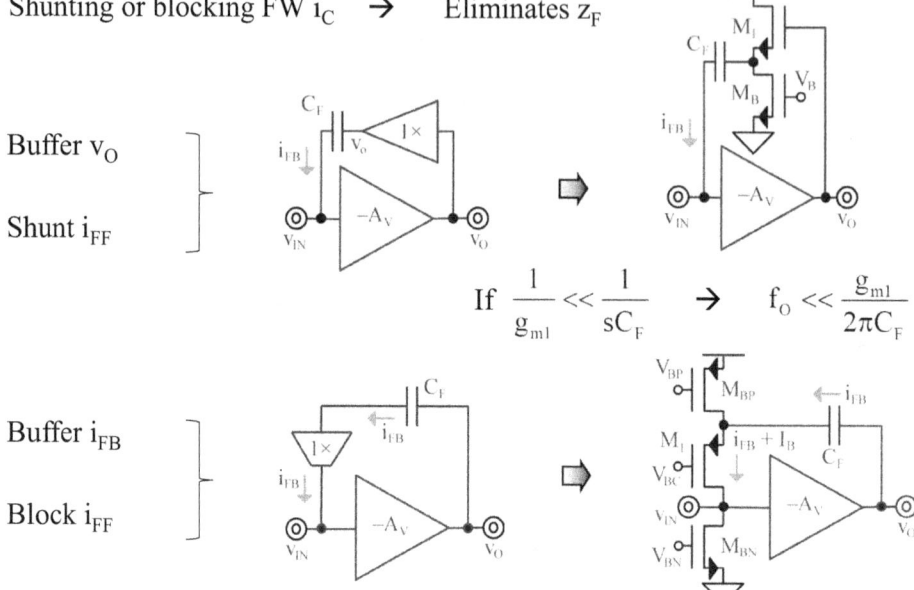

Buffer v_O

Shunt i_{FF}

If $\dfrac{1}{g_{m1}} << \dfrac{1}{sC_F}$ → $f_o << \dfrac{g_{m1}}{2\pi C_F}$

Buffer i_{FB}

Block i_{FF}

Example: Stabilize so PM = 45°

Solution:

Diode-connected Q_1 → 1 node ∴ 1 pole → Stable

Source-Degenerated M_O → 1 node ∴ 1 pole → Stable

Gate-looped M_O → 2 nodes ∴ 2 poles

f_I': $C_I \approx C_{\pi E}$ $R_I = r_{\pi E} \parallel r_{o2} \parallel R_{SO} \approx R_{SO} \approx \dfrac{2}{g_{mO}}$

f_G': $C_G \approx \left(\dfrac{r_{ds4}}{r_{\pi E}}\right) C_{GDO}$ $R_G = r_{oE} \parallel r_{ds3}$ → Can add C_S $A_{LG0} \equiv \dfrac{i_{fb}}{i_c} \approx g_{mE} R_G$

PM ≥ 45° when: $f_{0dB} \approx GBW = A_{LG0} p_G \approx \dfrac{g_{mE}}{2\pi C_S}$ $\approx \dfrac{-(r_{\pi E} \parallel r_{o2})(-g_{mE}) R_G g_{mO}}{1 + \left(g_{mO} + \dfrac{1}{r_{dsO}}\right)(r_{\pi E} \parallel r_{o2})}$

$p_G \approx \dfrac{1}{2\pi R_G C_S} << f_{0dB} \leq p_I \approx \dfrac{1}{2\pi R_{SO}\left[C_{\pi E} + (C_{GSO} \oplus C_S)\right]}$

E. Nested Cross-Amp (Miller) Stabilization

Three gain stages:

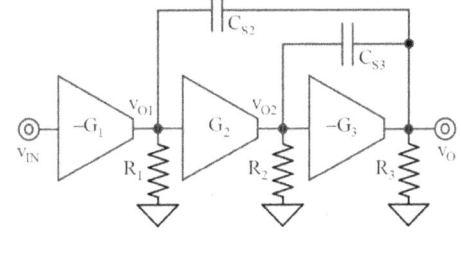

C_{S2} pulls p_1 to low f_O

    pushes p_O to high f_O

    introduces out-of-phase z_{S2} → Impede, block, shunt $i_{CS2(FW)}$

C_{S3} pulls p_2 to low f_O

    pushes p_O to high f_O

    introduces out-of-phase z_{S3} → Impede, block, shunt $i_{CS3(FW)}$

v_{O2}: G_3 shunt-samples v_{O2} & C_{S2} closes a – feedback loop about v_{O2}

    $\therefore$ Higher-f_O $R_{O2} \approx 1/G_{LG2}$ → Feedback pushes p_2 to high f_O

Example

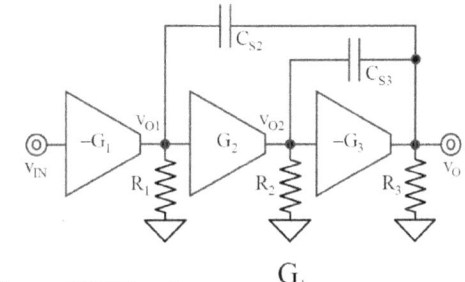

Target: $p_1 \ll f_{0dB} \approx p_2 \ll p_O$, z_{RHP}'s

$$A_{V0} = A_{V10}A_{V20}A_{V30} = (-G_1R_1)G_2R_2(-G_3R_3)$$

C_{S2I} shunts R_1: $p_1 \approx \dfrac{1}{2\pi R_1(-A_{V20}A_{V30}C_{S2})} \ll f_{0dB} \approx GBW = A_{V0}p_1 \approx \dfrac{G_1}{2\pi C_{S2}}$

C_{S2} loops v_{O2}:   Without R_{O1} $R_{O2} \approx \dfrac{1}{G_{LG2}} \approx \dfrac{1}{A_{V30}\left[C_{S2}/(C_{S2}+C_1)\right](-G_2)} \approx \dfrac{1}{-A_{V30}G_2}$

    $p_2 \approx \dfrac{1}{2\pi R_{O2}(C_2 + C_{S3I})} \approx \dfrac{1}{2\pi R_{O2}\left(-A_{V30}C_{S3}\right)} \approx \dfrac{G_2}{2\pi C_{S3}} \equiv f_{0dB}$

C_{S3} loops v_O: No R_{O1}, $R_{O2} \rightarrow$ Low A_{LGS2} $R_{O3} \approx \dfrac{1}{G_{LG3}} \approx \dfrac{1}{\left[C_{S3}/(C_{S3}+C_2)\right]G_3} \approx \dfrac{1}{G_3}$

    $p_O \approx \dfrac{1}{2\pi R_{O3}[(C_{S2} \oplus C_1) + (C_{S3} \oplus C_2) + C_{LD}]} \approx \dfrac{G_3}{2\pi C_{LD}} \gg f_{0dB}$

With bypass (feed-forward) zero: i_2 falls past p_1

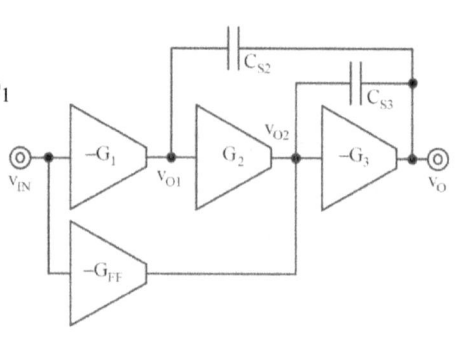

With $C_{O1} = C_1 + C_{S2I} \approx -A_{V20}A_{V30}C_{S2}$

Reversal z_{1X} appears when $i_2 \leq i_{FF}$:

$$i_2 = \left.\frac{v_{IN}A_{V10}G_2}{1+sR_1C_{O1}}\right|_{f_O \gg p_1} \approx \frac{v_{IN}(-G_1)G_2}{sC_{O1}} \leq i_{FF} = -v_{IN}G_{FF} \quad \rightarrow \quad z_{1X} \approx \frac{G_1G_2}{2\pi G_{FF}C_{O1}}$$

$$\frac{z_{1X}}{p_1} \approx \left(\frac{G_1G_2}{2\pi G_{FF}C_{O1}}\right)2\pi R_1 C_{O1} = \left(\frac{G_1}{G_{FF}}\right)\left(\frac{R_1}{1/G_2}\right) \quad \rightarrow \quad z_{1X} \text{ can be} \gg p_1$$

$$\frac{f_{0dB}}{z_{1X}} \approx \left(\frac{G_1}{2\pi C_{S2}}\right)\left(\frac{2\pi G_{FF}C_{O1}}{G_1G_2}\right) = \left(\frac{G_{FF}}{G_2}\right)\left(\frac{C_{O1}}{C_{S2}}\right) = -G_{FF}R_2A_{V30} \quad \rightarrow \quad z_{1X} \text{ can be} \ll f_{0dB}$$

Example: $p_1 < p_2 \approx z_{1X} < f_{0dB} \approx p_O \ll z_{RHP}\text{'s} \quad \rightarrow \quad$ Adjust/degenerate G_{FF}

Four gain stages:

C_{S2} pulls p_1 to low f_O

    pushes p_2 to high f_O

C_{S3} pulls p_1 to low f_O

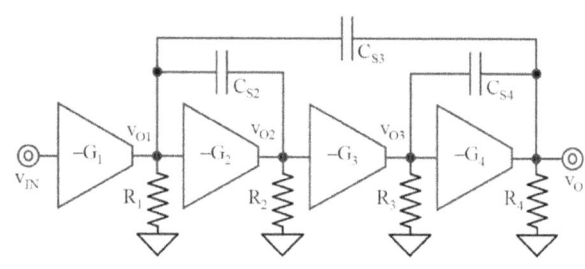

    pushes p_O to high f_O

C_{S4} pulls p_3 to low f_O

    pushes p_O to high f_O

v_{O3}: G_4 shunt-samples v_{O3} & C_{S3} closes a $-$ feedback loop about v_{O3}

    $\therefore$ Higher-f_O $R_{O3} \approx 1/G_{LG3} \quad \rightarrow \quad$ Feedback pushes p_3 to high f_O

$C_S\text{'s}$ introduce out-of-phase $z_S\text{'s} \quad \rightarrow \quad$ Null, block, shunt $i_{CS(FW)}\text{'s}$

Example: Target: $p_1 \ll f_{0dB} \approx p_3 \ll p_O, p_2, z_{RHP}$'s

$$A_{V0} = A_{V10}A_{V20}A_{V30}A_{V40}$$

C_{S31} shunts R_1: $\quad p_1 \approx \dfrac{1}{2\pi R_1(-A_{V20}A_{V30}A_{V40}C_{S3})} \ll f_{0dB} \approx GBW = A_{V0}p_1 \approx \dfrac{G_1}{2\pi C_{S3}}$

C_{S3} loops v_{O3}: No R_{O1} $\quad R_{O3} \approx \dfrac{1}{G_{LG3}} \approx \dfrac{1}{A_{V40}(-C_{S3}/C_{S2})G_3}$

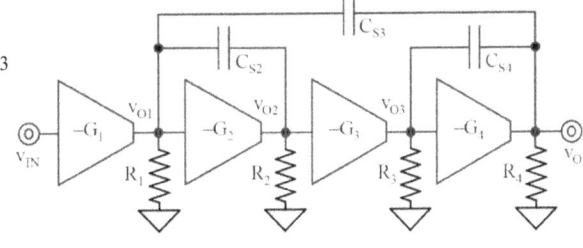

$$p_3 \approx \dfrac{1}{2\pi R_{O3}(C_3 + C_{S4I})} \approx \dfrac{1}{2\pi R_{O3}(-A_{V40}C_{S4})} \approx \left(\dfrac{G_3}{2\pi C_{S4}}\right)\left(\dfrac{C_{S3}}{C_{S2}}\right) \equiv f_{0dB}$$

C_{S2} loops v_{O2}:

Without R_{O1}, R_{O3} → Low A_{LGS3}

$$R_{O2} \approx \dfrac{1}{G_{LG2}} \approx \dfrac{C_{S2} + C_1 + C_{S3}}{C_{S2}G_2}$$

$$p_2 \approx \dfrac{1}{2\pi R_{O2}\{C_2 + [C_{S2} \oplus (C_1 + C_{S3})]\}} \approx \dfrac{1}{2\pi R_{O2}(C_{S2} \oplus C_{S3})} \approx \dfrac{G_2}{2\pi C_{S3}} \gg f_{0dB}$$

C_{S4} loops v_O: No R_{O1}, R_{O3}, A_{LGS3} $\quad R_{O4} \approx \dfrac{1}{G_{LG4}} \approx \dfrac{1}{\left[C_{S4}/(C_3 + C_{S4})\right]G_4} \approx \dfrac{1}{G_4}$

$$p_O \approx \dfrac{1}{2\pi R_{O4}\{[C_{S3} \oplus (C_1 + C_{S2I})] + (C_{S4} \oplus C_3) + C_{LD}\}} \approx \dfrac{G_4}{2\pi(C_{S3} + C_{LD})} \gg f_{0dB}$$

F. Design

1. From desired I/O translation:

    Consult reported literature

    Consult with colleagues

    Signal: All $s_{I/O}$ or   Alternating $s_{i/o}$

    Mix: Series   or   Shunt

    Sample: Series   or   Shunt

    Amp: A_G A_V A_I A_Z

2. From mixer & desirable R_I:

    High R_I $\therefore$ s_I to Gate/Base

    Low R_I $\therefore$ s_I to Source/Emitter

    ICMR can dictate N or P type

3. From sampler & desirable R_O:

    High R_O $\therefore$ s_O to Drain/Collector

    Low R_O $\therefore$ s_O to Source/Emitter

    v_O swing can dictate N or P type

4. From desired I/O translation:

    Feedback network

5. Amplify s_e to s_o

6. Bias circuit

7. Determine frequency response

8. Stabilize if necessary:

    Reach f_{0dB} with PM $\rightarrow$ $f_{0dB} < f_{180°}$

Example

Objective: High $i_{OUT(MAX)}$ High $v_{O(MAX)}$ Buffer v_I's alternating v_i

Design:

Series-mix v_i $\rightarrow$ With v_{GS} $\rightarrow$ $M_{E\beta}$

High R_I $\rightarrow$ v_I to Gate

Shunt-sample v_o $\rightarrow$ With Gate or Source

Low R_O $\rightarrow$ v_O to Source

High $v_{O(MAX)}$ $\rightarrow$ Use PFET

Translate v_o to v_{fb} $\rightarrow$ $v_o \approx v_{fb}$ $\therefore$ Use $M_{E\beta}$

High $i_{OUT(MAX)}$ at $v_{O(MAX)}$ $\rightarrow$ Loop to v_{DD}-supplied CS PFET $\rightarrow$ M_{A1}

Amplify v_e to M_{A1} $\rightarrow$ Fold/loop $i_{e\beta}$ into r_{ds}, M_{A1} $\rightarrow$ CB Q_{A2} into M_{A3}'s r_{ds3}

Bias $M_{E\beta}$, Q_{A2}, M_{A1} $\rightarrow$ Add M_{A4}, Q_{B1}, R_{B2}, M_{B3}, Mirror to M_{A3}:M_{B3}

Stabilize $\rightarrow$ p_{G1} = Low $\rightarrow$ Add C_s if needed to ensure $f_{0dB} \leq p_O \ll p_{E2}$

Chapter 4. Differential Stage

4.1. Differential Input

4.2. Common-Mode Noise

4.3. Power-Supply Noise

4.4. Electronic Noise

4.5. Mismatches

4.6. Summing Stage

4.7. Simulations

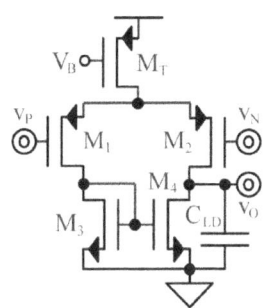

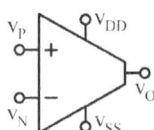

4.1. Differential Input: A. Large-Signal Operation

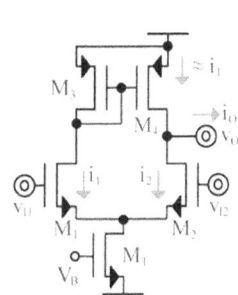

Aim: Double- to single-ended conversion

→   Combine differential i_1 & i_2 into $i_O = i_1 - i_2 = i_{OD}$

How: Mirror i_1 into i_2 & KCL i_O difference to v_O

Bias: – FB mixes v_{I1} & v_{I2} ∴ $V_{I1} \approx V_{I2}$

→ $V_{GS1} \approx V_{GS2}$ ∴ $I_1 \approx I_2 \approx 0.5I_T$

Input Common-Mode Range:

Low v_{IC} "crushes" M_T

→ $v_{IC} \geq v_{SS} + v_{DST(SAT)} + v_{TN12} + v_{DS12(SAT)}$

High v_{IC} "crushes" M_1 into M_3

→ $v_{IC} \leq v_{DD} - v_{SD3(SAT)} - |V_{TP0}| + v_{TN1}$

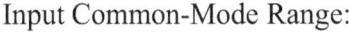

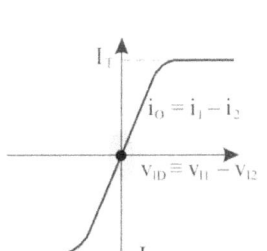

Example

Given: $v_{DD} = 2$ V, $i_T = 50$ μA, W_N's = 3 μm, W_P's = 10 μm, L's = 1 μm.

Process: $K_N' = 200$ μA/V^2, $K_P' = 40$ μA/V^2, $V_{TN0} = -V_{TP0} = 400$ mV, $L_{OL} = 30$ nm,

Solution: $L_{CH} = L - 2L_{OL} = 940$ nm $\qquad\qquad\qquad\qquad\qquad n_I = 2.$

$$V_{SDT(SAT)} \approx \sqrt{\frac{2i_T}{K_P'(W_P/L_{CH})}} = 480 \text{ mV}$$

$$V_{SD12(SAT)} \approx \sqrt{\frac{2(0.5i_T)}{K_P'(W_P/L_{CH})}} = 340 \text{ mV}$$

$$V_{DS34(SAT)} \approx \sqrt{\frac{2(0.5i_T)}{K_N'(W_N/L_{CH})}} = 280 \text{ mV}$$

$V_{SD/DS(SAT)}$'s $> 2n_I V_t \approx 100$ mV

∴ In Strong Inversion

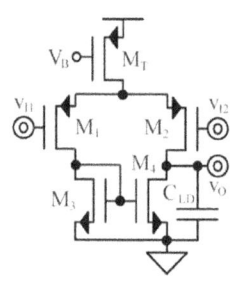

$$v_{IC} \leq v_{DD} - V_{SDT(SAT)} - |V_{TP0}| - V_{SD12(SAT)} = 780 \text{ mV}$$

$$v_{IC} \geq V_{TN0} + V_{DS3(SAT)} - |V_{TP0}| = 280 \text{ mV}$$

$$\text{Slew Rate} \equiv SR \equiv \text{Max}\left\{\frac{dv_O}{dt}\right\}$$

Often: $\qquad C_{LD} \gg C_{GS}$

$$\therefore \quad SR \approx \frac{i_{O(MAX/MIN)}}{C_O}$$

→ When $|v_{ID}| \gg v_{ID(MAX)}$

$$i_{O(MAX)} = i_{4(MAX)} - i_{2(MIN)} \approx i_T - 0 = i_T$$

$$i_{O(MIN)} = i_{4(MIN)} - i_{2(MAX)} \approx 0 - i_T = -i_T$$

$$SR \approx \pm\frac{i_T}{C_O} \quad \rightarrow \quad \text{Symmetrical}$$

52

B. Small-Signal Gain

Polarity: M_1, M_2, M_4 invert $\therefore$
$$\left[\begin{array}{l} v_{I1} = v_P = +0.5v_{id} \\[2mm] v_{I2} = v_N = -0.5v_{id} \end{array}\right.$$

Low-frequency Gain: $i_4 \approx i_1$

$$A_{D0} \equiv \frac{v_o}{v_{id}} = \frac{(i_4 - i_2)R_O}{v_{id}} \approx \left[\frac{0.5v_{id}g_{m1} - (-0.5v_{id}g_{m2})}{v_{id}}\right](r_{ds2} \| r_{ds4}) = g_{m12}R_O$$

p_1 OC Model: Split C_{GD}'s $\qquad$ $p_O \equiv$ Output Pole $\qquad$ $g_{m1} \approx g_{m2}$

If $f_O = \dfrac{1}{2\pi R_O C_O} \approx \dfrac{1}{2\pi(r_{ds2} \| r_{ds4})(C_{GD2} + C_{DB2} + C_{GD4} + C_{DB4} + C_{LD})} \approx \dfrac{1}{2\pi R_O C_{LD}} \ll f_{G3}$

$$f_{G3} = \frac{1}{2\pi R_{G3} C_{G3}} \approx \frac{g_{m3}}{2\pi\left\{C_{GD1} + C_{DB1} + C_{DB3} + C_{GS3} + C_{GS4} + C_{GD4}\left[1 + g_{m4}(r_{ds2} \| r_{ds4})\right]\right\}}$$

$\rightarrow\quad C_O$ shunts R_O past $p_1 = p_O \approx f_O \qquad$ If $p_O \ll$ Others $\therefore\quad f_{0dB} \approx GBW \approx \dfrac{g_{m12}}{2\pi C_O}$

p_2 SC Model: C_O shunts $R_O \quad\rightarrow\quad R_O$ fades $\quad\rightarrow\quad |A_{V4}| \ll g_{m4}R_O$

$$f_{G3}" = \frac{1}{2\pi R_{G3} C_{G3}"} \approx \frac{g_{m3}}{2\pi\left[C_{GD1} + C_{DB1} + C_{DB3} + C_{GS3} + C_{GS4} + (C_{GD4} \oplus C_{LD})\right]} \approx \frac{g_{m3}}{2\pi(2C_{GS34})}$$

$C_{G3}"$ shunts R_{G3} past $p_2 = p_M \approx f_{G3}" \rightarrow i_4 \propto \dfrac{1}{f_O} \qquad \therefore \quad p_M$ fades when i_2 overwhelms i_4

Reversal Zero z_{MX}

$G_D = 0.5g_{m1} + 0.5g_{m2}$ falls to $0.5g_{m2}$:

$$G_D \approx \frac{0.5g_{m1}}{1 + \dfrac{s}{2\pi p_M}} + 0.5g_{m2} = g_{m12}\left[\frac{1 + \dfrac{s}{2\pi(2p_M)}}{1 + \dfrac{s}{2\pi p_M}}\right] = g_{m12}\left(\frac{1 + \dfrac{s}{2\pi z_{MX}}}{1 + \dfrac{s}{2\pi p_M}}\right)$$

Insight: G_D falls $2\times$ to $0.5g_{m2}$ $\qquad$ Mirror Pole

$\therefore\quad f_O$ rises $2\times$ to $z_{MX} \rightarrow z_{MX} \approx 2p_M$

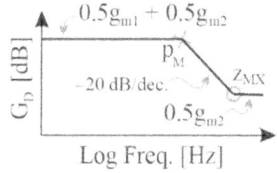

Example

Given: Same example with $C_{LD} = 100$ fF.

Additional Parameters: $\lambda_N = \lambda_P = 2\%$ when $L = 1$ μm, $t_{OX} = 5$ nm.

Small-Signal Parameters:

$$g_{m12} \approx \sqrt{2(0.5i_T)K_P'(W_P/L_{CH})} = 150 \ \mu S$$

$$g_{m34} \approx \sqrt{2(0.5i_T)K_N'(W_N/L_{CH})} = 180 \ \mu S$$

$$r_{ds12} = r_{ds34} \approx \frac{1}{(0.5i_T)\lambda_{N P}} = 2 \ M\Omega$$

$$C_{OX}" \approx \frac{\varepsilon_{Si}\varepsilon_0}{t_{OX}} = \frac{(3.9)(8.85p)}{5n} = 6.9 \ mF/m^2 = 6.9 \ fF/(\mu m)^2 = 6.9 \ fF/\mu m^2$$

$C_{GD12T} = C_{OL12T} = C_{OX}"W_PL_{OL} = 2.1$ fF $C_{GD34} = C_{OL34} = C_{OX}"W_NL_{OL} = 0.63$ fF

$C_{CH12T} = C_{OX}"W_P(L - 2L_{OL}) = 65$ fF   → $C_{GS12T} = C_{OL} + (2/3)C_{CH12T} = 45$ fF

$C_{CH34} = C_{OX}"W_N(L - 2L_{OL}) = 20$ fF   → $C_{GS34} = C_{OL} + (2/3)C_{CH34} = 14$ fF

Gain: $A_{D0} = g_{m12}(r_{ds2} \| r_{ds4}) = 150$ V/V $= 44$ dB $f_{0dB} \approx GBW \approx \dfrac{g_{m12}}{2\pi C_O} = 230$ MHz

$$A_{V40} = -g_{m4}(r_{ds2} \| r_{ds4}) = -180 \ V/V$$

$$p_O \approx f_O' \approx \frac{1}{2\pi(r_{ds2} \| r_{ds4})(C_{GD2} + C_{GD4} + C_{LD})} = 1.6 \ MHz$$

$$\ll f_{G3}' \approx \frac{g_{m3}}{2\pi\{C_{GD1} + C_{GS3} + C_{GS4} + C_{GD4}(1 - A_{V40})\}} = 200 \ MHz$$

$$p_M \approx \frac{g_{m3}}{2\pi\left[C_{GD1} + C_{GS3} + C_{GS4} + (C_{GD4} \oplus C_{LD})\right]} = 930 \ MHz \rightarrow z_{MX} \approx 2p_M = 1.9 \ GHz$$

54

4.2. Common-Mode Noise: A. Small-Signal Gain

Low-frequency Gain:

Mismatched Reflection of i_2 Mirror Translation

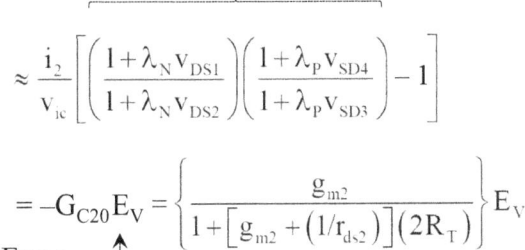

$$G_{C0} \equiv \frac{i_o}{v_{ic}} = \frac{i_4 - i_2}{v_{ic}} = \left(\frac{1}{v_{ic}}\right)\left[i_2\left(\frac{i_1}{i_2}\right)\left(\frac{i_4}{i_3}\right) - i_2\right]$$

Systemic v_D Mismatch

$$\approx \frac{i_2}{v_{ic}}\left[\left(\frac{1 + \lambda_N V_{DS1}}{1 + \lambda_N V_{DS2}}\right)\left(\frac{1 + \lambda_P V_{SD4}}{1 + \lambda_P V_{SD3}}\right) - 1\right]$$

v_T, K', & λ mismatches

add random error

$$= -G_{C20}E_V = \left\{\frac{g_{m2}}{1 + \left[g_{m2} + (1/r_{ds2})\right](2R_T)}\right\}E_V$$

% Error ⟶

→ Increase A_{C0}

$$A_{C0} \equiv \frac{v_o}{v_{ic}} = G_{C0}\left(r_{ds4} \| R_{D2}\right) \approx E_V\left(\frac{g_{m2}}{1 + 2R_I g_{m2}}\right)r_{ds4} \approx E_V\left(\frac{r_{ds4}}{2R_T}\right)$$

p_1 OC Model: Split C_{GD}'s C_O shunts R_O' past p_1' $= p_O$' $\approx f_O$'

If $f_O' = \dfrac{1}{2\pi R_O'C_O} \approx \dfrac{1}{2\pi r_{ds4}\left(C_{GD2} + C_{DB2} + C_{GD4} + C_{DB4} + C_{LD}\right)} \approx \dfrac{p_O}{2} \ll f_{G3}$

$$f_{G3} = \frac{1}{2\pi R_{G3}C_{G3}} \approx \frac{g_{m3}}{2\pi\left[C_{GD1} + C_{DB1} + C_{DB3} + C_{GS3} + C_{GS4} + C_{GD4}(1 + g_{m4}r_{ds4})\right]}$$

Reversal Pole

p_2 SC Model: C_O shunts R_O'   → R_O' fades

C_{G3} reduces i_4 ∴   Error $i_o = i_4 - i_2$ rises past z_M'

$i_o \approx -i_2$ past p_{MX}' when $i_4 \ll i_2$

$G_C = E_V G_{C20}$ rises by $\dfrac{1}{E_V}$ to G_{C20} from z_M' to p_{MX}'

Mirror Zero

∴ $p_{MX}' = \left(\dfrac{1}{E_V}\right)z_M'$ → $z_M' \approx E_V p_{MX}'$ C_{G3} shunts R_{G3} past $p_{MX}' = \dfrac{1}{2\pi R_{G3}C_{G3}} = p_M$

55

G_{C2}: $0.5C_T$ shunts $2R_T$ degeneration → G_{C2} rises past z_D

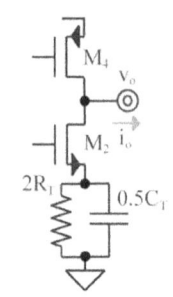

$$\frac{1}{s(0.5C_T)} \leq 2R_T \quad \rightarrow \quad f_o \geq \frac{1}{2\pi R_T C_T} \approx z_D$$

$$\downarrow$$

Degeneration Zero

Until $0.5C_T$ shorts $2R_T \parallel R_{S2}$ → G_{C2} reaches g_{m2} past p_{DX}

C_O shunts R_O' before p_{DX}

$$\frac{1}{s(0.5C_T)} \leq 2R_T \parallel R_{S2} \approx \frac{1}{g_{m2}} \quad \rightarrow \quad f_o \geq \frac{g_{m2}}{2\pi(0.5C_T)} \approx p_{DX} \rightarrow \text{Reversal Pole}$$

G_{C2} rises by $1 + 2R_T g_{m2}$ from z_D to p_{DX}

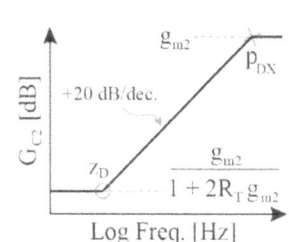

$$\therefore \quad p_{DX} \approx (1 + 2R_T g_{m2})z_D \approx 2R_T g_{m2} z_D$$

Example

Given: Same when $V_O \approx V_{IC} \approx 450$ mV

$$r_{dsT} = \frac{1}{\lambda_p i_T} = 1 \ M\Omega$$

Solution: $V_{S12} = V_{IC} + |V_{TP0}| + V_{SD12(SAT)} = 1.2$ V

$$V_{G34} = V_{TN0} + V_{DS34(SAT)} = 680 \text{ mV}$$

$$z_M' \approx E_V p_{MX}' = 8.3 \text{ MHz}$$

$$E_V = \left| \left(\frac{1 + \lambda_p V_{SD1}}{1 + \lambda_p V_{SD2}} \right) \left(\frac{1 + \lambda_N V_{DS4}}{1 + \lambda_N V_{DS3}} \right) - 1 \right| = \left| \left[\frac{1 + \lambda_p (V_{S12} - V_{G34})}{1 + \lambda_p (V_{S12} - V_O)} \right] \left(\frac{1 + \lambda_N V_O}{1 + \lambda_N V_{G34}} \right) - 1 \right| = 0.94\%$$

$$A_{C0} \approx E_V \left(\frac{g_{m2}}{1 + 2r_{dsT} g_{m2}} \right) r_{ds4} = 0.94\% = -41 \text{ dB} \qquad\qquad p_O' < z_M' < z_D < p_{MX}' < p_{DX}$$

$$p_O' \approx \frac{1}{2\pi R_O' C_O} \approx \frac{1}{2\pi r_{ds4}(C_{GD2} + C_{GD4} + C_{LD})} = 780 \text{ kHz} \approx \frac{p_O}{2}$$

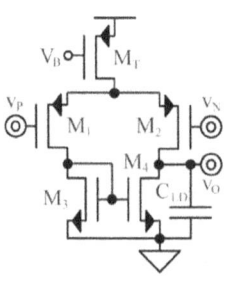

$$p_{MX}' \approx \frac{g_{m3}}{2\pi \left[C_{GD1} + C_{GS3} + C_{GS4} + (C_{GD4} \oplus C_O) \right]} = 930 \text{ MHz} \approx p_M$$

$$z_D \approx \frac{1}{2\pi r_{dsT} C_{GDT}} = 76 \text{ MHz} \qquad\qquad p_{DX} \approx \frac{2g_{m2}}{2\pi C_{GDT}} = 23 \text{ GHz}$$

B. Common-Mode Rejection Ratio

CMRR ≡ How much circuit favors v_{ID} input over v_{IC} noise

$$\equiv \frac{A_D}{A_C} = \left[\frac{A_{D0}\left(1+\dfrac{s}{2\pi z_{MX}}\right)}{\left(1+\dfrac{s}{2\pi p_O}\right)\left(1+\dfrac{s}{2\pi p_M}\right)}\right]\left[\frac{\left(1+\dfrac{s}{2\pi p_O'}\right)\left(1+\dfrac{s}{2\pi p_{MX}'}\right)\left(1+\dfrac{s}{2\pi p_{DX}}\right)}{A_{C0}\left(1+\dfrac{s}{2\pi z_M'}\right)\left(1+\dfrac{s}{2\pi z_D}\right)}\right]$$

$$\approx \frac{A_{D0}\left(1+\dfrac{s}{2\pi z_{MX}}\right)\left(1+\dfrac{s}{2\pi p_{DX}}\right)}{A_{C0}\left(1+\dfrac{s}{2\pi z_M'}\right)\left(1+\dfrac{s}{2\pi z_D}\right)} \approx \frac{g_{m12}\left(r_{ds2}\|r_{ds4}\right)\left(1+\dfrac{s}{2\pi z_{MX}}\right)\left(1+\dfrac{s}{2\pi p_{DX}}\right)}{\dfrac{E_V g_{m12} r_{ds4}}{1+2R_T g_{m12}}\left(1+\dfrac{s}{2\pi z_M'}\right)\left(1+\dfrac{s}{2\pi z_D}\right)}$$

$$\approx \left[\frac{2R_T g_{m12} r_{ds2}}{E_V\left(r_{ds2}+r_{ds4}\right)}\right]\left[\frac{\left(1+\dfrac{s}{2\pi z_{MX}}\right)\left(1+\dfrac{s}{2\pi p_{DX}}\right)}{\left(1+\dfrac{s}{2\pi z_M'}\right)\left(1+\dfrac{s}{2\pi z_D}\right)}\right]$$

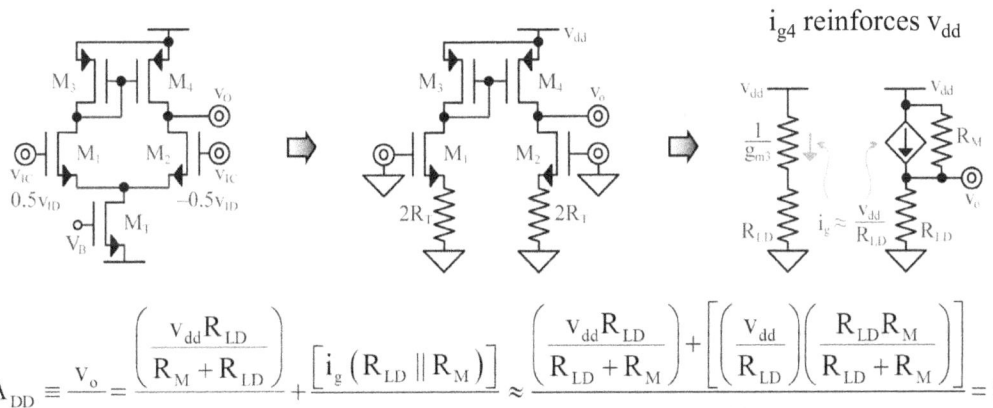

4.3. Power-Supply Noise: A. P-Type Mirror Load: i. Positive-Supply Gain

Positive-supply noise v_{dd} = Small signal ≠ 0

Small Signals ∴ Exclude V_{ID} V_{IC} V_{DD} V_{SS} I_T

Superposition ∴ $v_{id} \equiv v_{ic} \equiv v_{ss} \equiv i_t \equiv 0$ R_M couples v_{dd}

i_{g4} reinforces v_{dd}

$$A_{DD} \equiv \frac{v_o}{v_{dd}} = \frac{\left(\dfrac{v_{dd}R_{LD}}{R_M+R_{LD}}\right)}{v_{dd}} + \frac{\left[i_g\left(R_{LD}\|R_M\right)\right]}{v_{dd}} \approx \frac{\left(\dfrac{v_{dd}R_{LD}}{R_{LD}+R_M}\right)+\left[\left(\dfrac{v_{dd}}{R_{LD}}\right)\left(\dfrac{R_{LD}R_M}{R_{LD}+R_M}\right)\right]}{v_{dd}} = 1$$

P-mirrors with balanced loads reproduce v_{dd} Approximation: $R_{LD} \gg \dfrac{1}{g_{m3}}$

ii. Negative-Supply Gain

Negative-supply noise v_{ss} = Small signal $\neq$ 0

Small Signals $\quad \therefore \quad$ Exclude $\quad V_{ID} \quad V_{IC} \quad V_{DD} \quad V_{SS} \quad I_T$

Superposition $\quad \therefore \quad v_{id} \equiv v_{ic} \equiv v_{dd} \equiv i_t \equiv 0$ $\qquad\qquad R_{LD}'$ couples v_{ss}

$\qquad\qquad\qquad\qquad\qquad\qquad\qquad\qquad\qquad\qquad\qquad\qquad\qquad\qquad\qquad$ i_{g4} removes v_{ss}

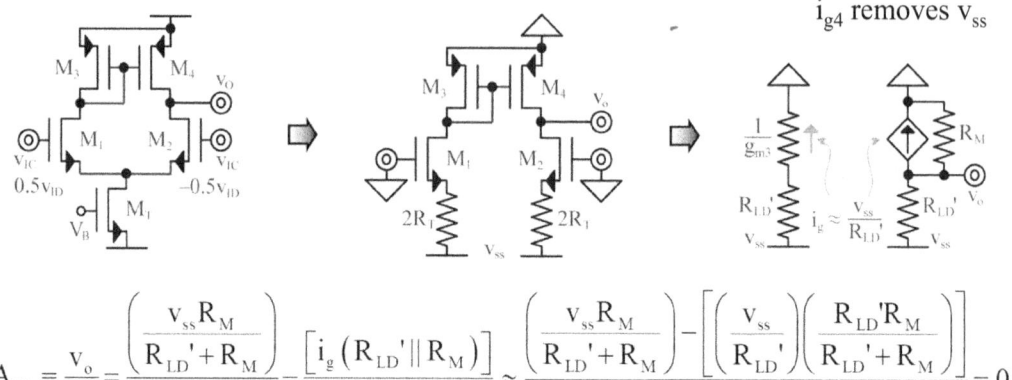

$$A_{SS} \equiv \frac{v_o}{v_{ss}} = \frac{\left(\dfrac{v_{ss}R_M}{R_{LD}'+R_M}\right)}{v_{ss}} - \frac{\left[i_g\left(R_{LD}'\|R_M\right)\right]}{v_{ss}} \approx \frac{\left(\dfrac{v_{ss}R_M}{R_{LD}'+R_M}\right) - \left[\left(\dfrac{v_{ss}}{R_{LD}'}\right)\left(\dfrac{R_{LD}'R_M}{R_{LD}'+R_M}\right)\right]}{v_{ss}} = 0$$

P-mirrors with balanced loads cancel v_{ss} $\qquad\longrightarrow\qquad$ Approximation: $R_{LD}' \gg \dfrac{1}{g_{m3}}$

B. N-Type Mirror Load

Supply noise $v_{dd/ss}$ = Small signal $\neq$ 0

Small Signals $\quad \therefore \quad$ Exclude $\quad V_{ID} \quad V_{IC} \quad V_{DD} \quad V_{SS} \quad I_T$

Superposition $\quad \therefore \quad v_{id} \equiv v_{ic} \equiv i_t \equiv 0$

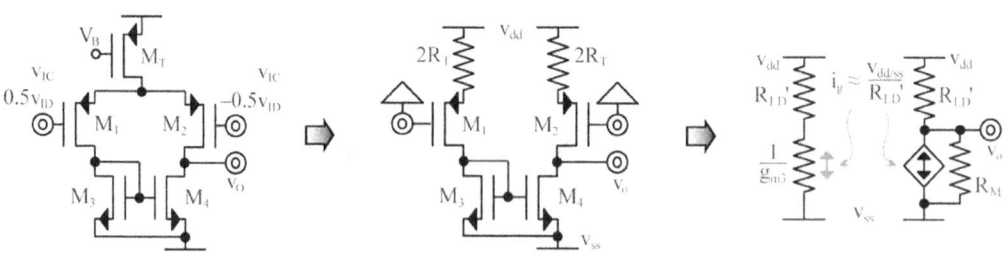

N-mirrors with balanced loads:

v_{dd}:	R_{LD}' couples v_{dd}	i_{g4} removes v_{dd}	$\therefore$	Cancel v_{dd}
v_{ss}:	R_M couples v_{ss}	i_{g4} reinforces v_{ss}	$\therefore$	Reproduce v_{ss}

Example

Given: Same when $V_O \approx V_{G3} \approx 680$ mV $\approx V_{DS34}$

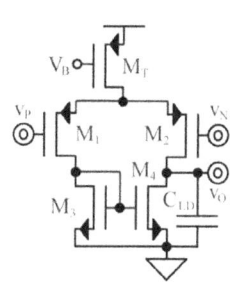

Analysis: $A_{DD} \approx \dfrac{\left(\dfrac{v_{dd}}{R_{T2}}\right) r_{ds4}}{v_{dd}} - \dfrac{\left(\dfrac{v_{dd}}{R_{T1}}\right)\left(R_{D2} \| r_{ds4}\right)}{v_{dd}} = -84$ dB

$$R_{T1} \approx 2r_{dsT} + \frac{r_{ds1} + 1/g_{m3}}{1 + g_{m1}r_{ds1}} = 2.007 \text{ M}\Omega \qquad R_{T2} \approx 2r_{dsT} + \frac{r_{ds2} + r_{ds4}}{1 + g_{m2}r_{ds2}} = 2.014 \text{ M}\Omega$$

$$R_{D12} = r_{ds12} + 2r_{dsT} + g_{m12}r_{ds12}(2r_{dsT}) = 604 \text{ M}\Omega$$

M_{34}'s v_D mismatch

distorts mirror translation

$$A_{SS} \approx \frac{\left(\dfrac{v_{ss}R_{D2}}{r_{ds4} + R_{D2}}\right)}{v_{ss}} + \frac{\left[\left(\dfrac{v_{ss}}{1/g_{m3} + R_{D1}}\right)\left(R_{D2} \| r_{ds4}\right)\right]}{v_{ss}} = 1$$

$\therefore \quad A_{DD} > -84$ dB

$A_{SS} \neq$ Exactly 1 V/V

C. Power-Supply Rejection Ratio

Power-supply rejection $\equiv$ Inability to amplify $v_{dd/ss}$ $\rightarrow$ $PSR \equiv \dfrac{1}{A_{DD/SS}} \equiv \dfrac{\partial v_{DD\cdot SS}}{\partial s_O}$

PSRR $\equiv$ How much circuit favors

$\quad v_{ID}$ input over $v_{DD/SS}$ noise

$$PSRR \equiv \frac{A_D}{A_{DD\cdot SS}} = A_D PSR$$

Balanced P-type mirrors reproduce v_{dd} $\qquad \therefore$ $\qquad PSRR^+ \equiv \dfrac{A_D}{A_{DD}} \approx \dfrac{A_D}{1} \approx A_D$

$\qquad\qquad\qquad$ cancel v_{ss} $\qquad \therefore$ $\qquad PSRR^- \equiv \dfrac{A_D}{A_{SS}} \approx \dfrac{A_D}{0} \rightarrow \infty$

Balanced N-type mirrors reproduce v_{ss} $\qquad \therefore$ $\qquad PSRR^- \equiv \dfrac{A_D}{A_{SS}} \approx \dfrac{A_D}{1} \approx A_D$

$\qquad\qquad\qquad$ cancel v_{dd} $\qquad \therefore$ $\qquad PSRR^+ \equiv \dfrac{A_D}{A_{DD}} \approx \dfrac{A_D}{0} \rightarrow \infty$

4.4. Electronic Noise: A. Model

Base/gate-referred noise $v_{b/g}^*$ ≡ Derived from measured non-degenerated noise i_n^*

Analysis: $s_x^* \ll S_X$ ∴ Use small-signal translations

$$v_{b/g}^* \equiv \frac{i_n^*}{g_m} \;\rightarrow\; i_{c/d}^* \approx G_M \text{ Translation of } v_{b/g}^*$$

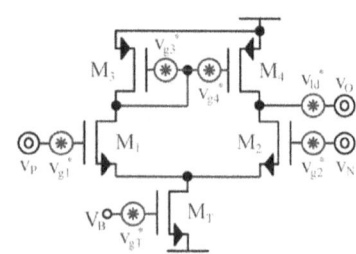

E/S-Degenerated Transistors:

$$i_{c/d}^* = v_{b/g}^* G_M \approx \left(\frac{i_n^*}{g_m}\right)\left(\frac{g_m}{1+g_m R_S}\right) = \frac{i_n^*}{1+g_m R_S} \;\rightarrow\; R_{E/S} \text{ deg. suppresses } i_{c/d}^*$$

Multiple Noise Sources: Random & uncorrelated ∴ Statistical sum

$$s_{total}^* = \sqrt{\left(s_{n1}^*\right)^2 + \left(s_{n2}^*\right)^2 + \left(s_{n3}^*\right)^2 + ... + \left(s_{n(k)}^*\right)^2}$$

→ Root–sum–squared (RSS) magnifies dominant terms (i.e., others fade)

Design Note: Vertical BJT, sub-surface JFET i_n < PMOS i_n^* < NMOS i_n^*

B. Input-Referred Noise

Contributions: All components generate electronic noise

Effect of Noise: Electronic noise propagates to s_O

Input-Referred Metric: s_{in}^* = Level above which $s_{IN}A_{IO}$ overcomes s_o^* in s_O

Analog Requirement: $s_o^* \ll s_O$ ∴ $s_{in}^* = \dfrac{s_o^*}{A_{IO}} \ll s_{IN} = \dfrac{s_O}{A_{IO}}$

Common Mode: i_t^* splits ∴ $i_o^* \approx i_1 - i_2 \approx 0.5 i_t^* - 0.5 i_t^* = 0$

Input-Referred Analysis: A_D gain suppresses effect of v_{Id}^*

$$v_{id}^* \approx \sqrt{2\left(\frac{i_{12}^*}{g_{m12}}\right)^2 + 2\left(\frac{i_{34}^*}{g_{m12}}\right)^2 + \left(\frac{v_{Id}^*}{A_D}\right)^2}$$

Signal-to-noise ratio: $SNR \equiv \dfrac{\Delta v_{ID}}{v_{id}^*}$

4.5. Mismatches: A. Analysis

Random: $\Delta i_{c/d}^* \equiv$ Statistical mismatch of non-degenerated $i_{C/D}$'s $\approx 5\%-15\%$

Systemic: $\Delta i_{c/d} \equiv$ Voltage-produced ($\Delta v_{C/D}$) mismatch of $i_{C/D}$'s

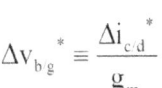

Analysis: $\Delta i_{c/d}^*$ & $\Delta i_{c/d} \ll I_{C/D}$ ∴ Use small-signal translations

Offset: $I_{OS} = I_{OS(S)} \pm I_{OS}^* =$ Linear $\pm$ Root–Sum–Squared (RSS)

E/S Degenerated Transistors:

$$\Delta v_{b/g}^* \equiv \frac{\Delta i_{c/d}^*}{g_m}$$

$$I_{OS(S)} = i_1 - i_2 = \frac{v_D + \Delta v_D}{R_{D1}} - \frac{v_D}{R_{D2}} = \frac{\Delta v_D}{r_{ds12} + R_S + g_{m12} r_{ds12} R_S}$$

$$I_{OS}^* = \Delta i_{12}^* = \Delta v_g^* G_M \approx \left(\frac{\Delta i_d^*}{g_{m12}}\right)\left(\frac{g_{m12}}{1 + g_{m12} R_S}\right) = \frac{\%I_{12}}{1 + g_{m12} R_S}$$

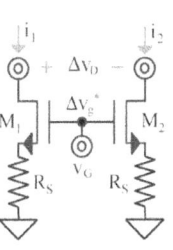

→ R_S degeneration suppresses effects of $\Delta v_{C/D}$ & $\Delta i_{c/d}^*$

B. Input-Referred Offset

Mismatches:

$M_1 : M_2$ → $\Delta v_{b/g12}^*$ ⎫

$M_3 : M_4$ → $\Delta v_{b/g34}^*$ ⎬ Random

$M_B : M_T$ → $\Delta v_{b/gBT}^*$ ⎭

$v_{D13} : v_{D24}$ → Δv_D ⎬ Systemic → Zero-offset $v_O = v_{DD} - v_{SG3}$

Common Mode: Δi_{BT}^* splits ∴ $i_o^* = i_1 - i_2 \approx 0.5\Delta i_{BT}^* - 0.5\Delta i_{BT}^* = 0$

Random Offset: Each Δi_x^* offsets i_O ∴ $V_{OS}^* = \sqrt{\left(\frac{\Delta i_{12}^*}{g_{m12}}\right)^2 + \left(\frac{\Delta i_{34}^*}{g_{m12}}\right)^2}$

Systemic Offset: $V_{OS(S)} = \frac{\Delta v_D}{A_{D0}}$ ∴ Higher A_{D0} desensitizes stage $\quad n_I > 1$
$\downarrow$

Design Note: BJTs' exp g_m suppresses Δi_x^* more than FETs' exp/quadratic g_m

Example

Given: Same when $V_O \approx 450$ mV, $\Delta i_{MOS}{}^* = \pm 5\%$

Analysis:

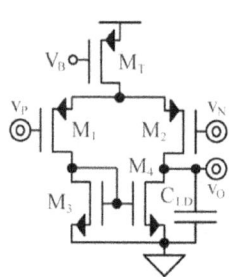

$$\Delta v_D = v_O - v_{G34} = -230 \text{ mV}$$

$$V_{OS(S)} = \frac{\Delta v_D}{g_{m12}(r_{ds2} \| r_{ds4})} = -1.5 \text{ mV}$$

$$V_{OS}{}^* = \sqrt{\left(\frac{\Delta i_{12}{}^*}{g_{m12}}\right)^2 + \left(\frac{\Delta i_{34}{}^*}{g_{m12}}\right)^2} = \sqrt{2\left[\frac{5\%(0.5 i_T)}{g_{m12}}\right]^2} \approx 12 \text{ mV}$$

$$V_{OS} = V_{OS(S)} \pm V_{OS}{}^* \approx -1.5 \pm 12 \text{ mV} \quad \rightarrow \quad \text{Statistical } 3\sigma \text{ Spread}$$

4.6. Summing Stage

Concept: g_m currents combine $\therefore$ v_{ID} translations add

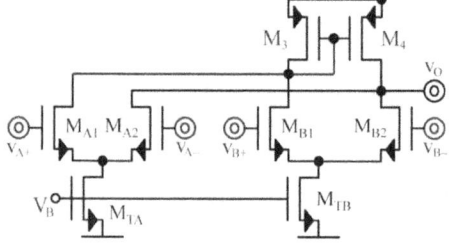

$$i_4 \approx i_1's = v_{A+}g_{mA} + v_{B+}g_{mB}$$

$$i_2's = v_{A-}g_{mA} + v_{B-}g_{mB}$$

$$i_O \approx i_1's - i_2's \approx v_A g_{mA} + v_B g_{mB}$$

If $g_{mA} = g_{mB} \equiv A_G$:

$$i_O = (v_{A+} - v_{A-} + v_{B+} - v_{B-})A_G \qquad = (v_{A+} + v_B - v_{A-})A_G \qquad = (v_A + v_B)A_G$$

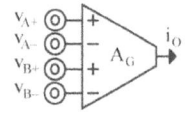   Offset   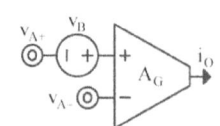   Analog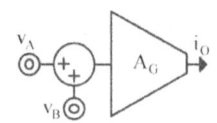

Translation   Summer

Programmable offset: $v_{OS} = v_B k_X = v_B \left(\dfrac{g_{mB}}{g_{mA}}\right)$

4.7. Simulations: A. Methods

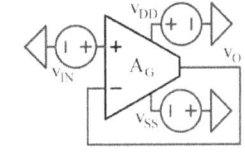

Bias:	Unity-gain feedback
$v_{IC/O}$ Range:	Bias & sweep v_{IN}
	Monitor $\quad i_T \quad v_O \approx v_{IN}$

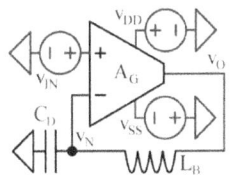

A_D:	Bias without closing loop $\rightarrow$ L_B
	Remove v_{ac} from v_N $\rightarrow$ C_D
	Inject $v_{ac} = 1$ into v_P $\rightarrow$ With v_{IN}
	Monitor $v_o = (v_p - v_n)A_D \approx (1 - 0)A_D = A_D$
A_C:	Bias & drive like A_D & ac-short v_P to v_N $\rightarrow$ C_C
$A_{DD/SS}$:	Like A_D, but inject $v_{ac} = 1$ into $v_{DD/SS}$ $\rightarrow$ With $v_{DD/SS}$

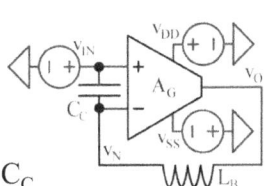

B. Example: Same

```
* PMOS Diff. Stage
vdd vdd 0 dc=2 ac=0
vss vss 0 dc=0 ac=0
ib vb vss dc=50u
mb vb vb vdd vdd pmos1 w=10u l=1u
mt vs vb vdd vdd pmos1 w=10u l=1u
m1 vg vp vs vs pmos1 w=10u l=1u
m2 vo vn vs vs pmos1 w=10u l=1u
m3 vg vg vss vss nmos1 w=3u l=1u
m4 vo vg vss vss nmos1 w=3u l=1u
cld vo vss 100f
vin vp 0 dc=450m ac=1 sin 450m 50m 1e6
*vin vp 0 dc=690m
.model nmos1 nmos vto=0.4 kp=200u ld=30n lambda=20m cgso=207p cgdo=207p
.model pmos1 pmos vto=-0.4 kp=40u ld=30n lambda=20m cgso=207p cgdo=207p
* tox=5n changes 2nd-order effects that alter results.
```

```
* Derived cgs model:
cgs34 vg vss 28f
cgs1 vp vs 45f
cgs2 vn vs 45f
.op
rb vo vn 1m
*.dc vin 0 2 1m
*.ac dec 100 10k 10e9
*lb vo vn 10k
*cd vn 0 10k
*cc vn vp 10k
*.tran 5u
.end
```

Chapter 5. Operational Amplifiers

5.1. Introduction

5.2. Input

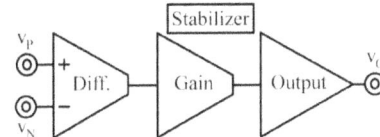

5.3. Output

5.4. Class-A Op-Amp Example

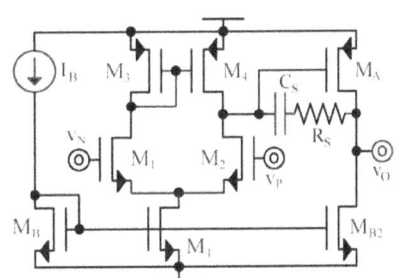

5.5. Class-AB Op-Amp Examples

5.6. Current-Mode Op Amp

5.1. Introduction

Applications: Filters

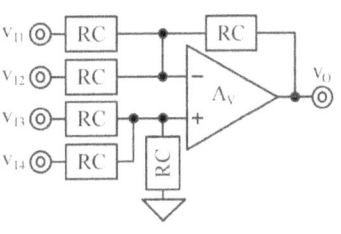

Mixers

Controllers   ...

Composition:

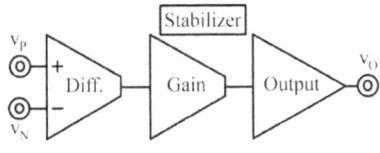

v_{ID} Range	v_O Range	A_D	$v_{DD} - v_{SS}$ Range
v_{IC} Range	i_O Range	f_{0dB}	T_J Range
R_{ID}	R_O	A_C	$P_{DD/SS}$
V_{OS}	SR	$A_{DD/SS}$	Cost
		v_{in}^*	...

5.2. Input: A. Mirror Fold

Differential Stage $\equiv$ N/P differential pair into P/N mirror

Modification: Fold N/P differential pair into N/P mirror with current mirrors

Similarities: v_{IC} Range

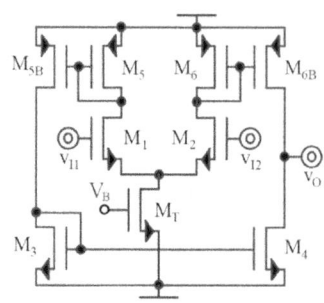

M_3:M_4's p_M & z_{MX}

Differences (relative to differential stage):

$$v_{SS} + v_{DS4(SAT)} \leq v_O \leq v_{DD} - v_{SD6B(SAT)} \equiv \text{Rail-to-Rail}$$

$$i_{O(MAX/MIN)} = \pm i_T \left(\frac{S_{6B}}{S_6} \right) \equiv \pm i_T \left(\frac{S_{5B}}{S_5} \right) \left(\frac{S_4}{S_3} \right) \left. \right\}$$
Bias balances when
mirror translations match

$$A_D = G_D R_O \approx \left[\left(\frac{g_{m2}}{2} \right) \left(\frac{S_{6B}}{S_6} \right) - \left(-\frac{g_{m1}}{2} \right) \left(\frac{S_{5B}}{S_5} \right) \left(\frac{S_4}{S_3} \right) \right] R_O \approx g_{m12} \left(\frac{S_{6B}}{S_6} \right) (r_{ds4} \parallel r_{ds6B})$$

$$p_O = \frac{1}{2\pi R_O C_O} \approx \frac{1}{2\pi (r_{ds4} \parallel r_{ds6B}) C_{LD}} \quad \rightarrow \quad \text{Similar } p_O$$

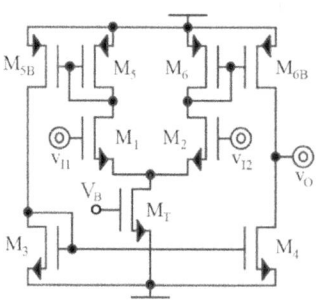

C_{G5} & C_{G6} reduce i_1 & i_2 projections to i_{5B} & i_{6B}:

$$G_D = \frac{i_{6B} - i_4}{v_{ID}} \approx \frac{0.5 G_{D0}}{1 + s/2\pi p_{G6}} + \frac{0.5 G_{D0}}{1 + s/2\pi p_{G5}} = \frac{G_{D0}}{1 + s/2\pi p_F}$$

$$p_F = \frac{1}{2\pi (r_{ds12} \parallel r_{ds56} \parallel R_{G56}) C_{G56}} \approx \frac{g_{m56}}{2\pi (C_{GS56} + C_{GS56B})} \quad \rightarrow \quad \text{1 High-}f_O \text{ Folding Pole}$$

$v_{D1} \approx v_{D2} \rightarrow$ Good M_1:M_2 reflection $\quad \underline{\quad} p_O$ reduces A_{V6B}

$$A_{C0} = G_{C0} R_O \approx G_{C120} \left[\left(\frac{1 + \lambda v_{SD6B}}{1 + \lambda v_{SD6}} \right) - \left(\frac{1 + \lambda v_{SD5B}}{1 + \lambda v_{SD5}} \right) \left(\frac{1 + \lambda v_{DS4}}{1 + \lambda v_{DS3}} \right) \right] \left(\frac{S_{6B}}{S_6} \right) R_O$$

$$= \left(\frac{g_{m12}}{1 + 2 r_{dsT} g_{m12}} \right) E_V \left(\frac{S_{6B}}{S_6} \right) (r_{ds4} \parallel r_{ds6B}) \quad \rightarrow \quad \text{More mirrors} \quad \therefore \quad \text{Higher } E_V \text{ \& } A_C$$

Balanced N-type M_3:M_4 mirror:

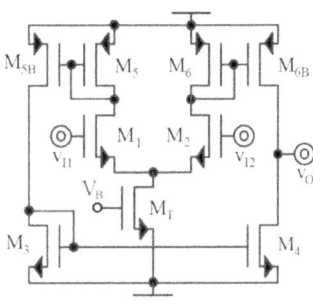

$$\therefore \quad A_{DD} \approx 0 \text{ V/V} \qquad A_{SS} \approx 1 \text{ V/V}$$

More non-degenerated pairs than in diff. stage:

$$\therefore \quad \text{Higher } v_{in}^*$$

$$v_{in}^* \approx \sqrt{2\left(\frac{i_{12}^*}{g_{m12}}\right)^2 + 2\left(\frac{i_{56}^*}{g_{m12}}\right)^2 + \left(\frac{i_{5B}^*}{g_{m1}}\frac{S_5}{S_{5B}}\right)^2 + \left(\frac{i_3^*}{g_{m1}}\frac{S_5}{S_{5B}}\right)^2 + \left(\frac{i_4^*}{g_{m2}}\frac{S_6}{S_{6B}}\right)^2 + \left(\frac{i_{6B}^*}{g_{m2}}\frac{S_6}{S_{6B}}\right)^2}$$

$$\therefore \quad \text{Higher } V_{OS} \;\rightarrow\; V_{OS(S)} = \frac{\Delta v_{D34}}{A_D} + \frac{\Delta v_{D6}}{A_D} + \left(\frac{\Delta v_{D5}}{r_{ds5B}}\right)\left(\frac{S_5}{S_{5B}}\right)\left(\frac{1}{g_{m1}}\right)$$

$$v_{D1} \approx v_{D2}$$

$$V_{OS}^* \approx \sqrt{\left(\frac{\Delta i_{12}^*}{g_{m12}}\right)^2 + \left(\frac{\Delta i_{55B}^*}{g_{m1}}\right)^2 + \left(\frac{\Delta i_{34}^*}{g_{m1}}\frac{S_5}{S_{5B}}\right)^2 + \left(\frac{\Delta i_{66B}^*}{g_{m2}}\right)^2}$$

B. Cascode Fold

Modification: Fold with cascodes

Architecture: Steer i_1 & i_2 with CB/G current buffers

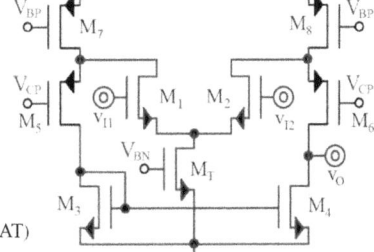

Similarities: $v_{IC(MIN)}$ SR (same $\pm i_{O(MAX)}$)

M_3:M_4's p_M & z_{MX}

Differences: Design: $v_{SD78} = v_{DD} - v_{D12} \geq v_{SD78(SAT)}$

$$v_{IC} \leq v_{D12} + v_{TN12} = V_{CP} + v_{SG56} + v_{TN12} = V_{CP} + |v_{TP56}| + v_{SD56(SAT)} + v_{TN12}$$

$$v_{SS} + v_{DS4(SAT)} \leq v_O \leq v_{D2} - v_{SD6(SAT)} = V_{CP} + |v_{TP56}| + v_{SD56(SAT)} - v_{SD6(SAT)}$$

$$A_D = G_D R_O = [0.5 g_{m1} - (-0.5 g_{m2})](R_{D6} \parallel r_{ds4}) \approx g_{m12} r_{ds4}$$

$$p_O = \frac{1}{2\pi R_O C_O} = \frac{1}{2\pi\left(r_{ds4} \parallel R_{D6}\right)C_O} \approx \frac{1}{2\pi r_{ds4} C_{LD}} \quad \rightarrow \quad \text{Slightly lower } p_O$$

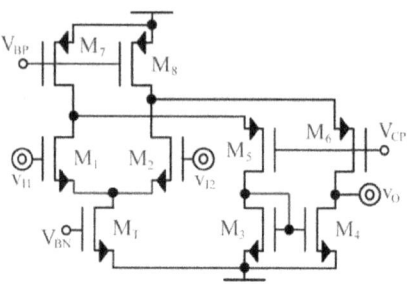

C_{S5} & C_{S6} reduce i_1 & i_2 projections to i_3 & i_O:

$$G_D = \frac{i_6 - i_4}{v_{ID}} \approx \frac{0.5 G_{D0}}{1 + s/2\pi p_{S6}} + \frac{0.5 G_{D0}}{1 + s/2\pi p_{S5}} \approx \frac{G_{D0}}{1 + s/2\pi p_F}$$

C_O shunts R_O past p_O

$$p_F = \frac{1}{2\pi R_{S56} C_{S56}} = \frac{1}{2\pi\left(r_{ds12} \parallel r_{ds78} \parallel R_{S56}\right)C_{S56}} \approx \frac{g_{m56}}{2\pi C_{GS56}} \quad \rightarrow \quad \text{High-}f_O \text{ Folding Pole}$$

$$A_{C0} = G_{C0} R_O \approx G_{C2}\left[\left(\frac{1 + \lambda v_{DS4}}{1 + \lambda v_{DS3}}\right) - 1\right] R_O \approx \left(\frac{g_{m2}}{1 + 2 r_{dsT} g_{m2}}\right)\overbrace{\left[\left(\frac{1 + \lambda v_O}{1 + \lambda v_{GS3}}\right) - 1\right]}^{E_V} r_{ds4}$$

Deg. M_{56} $v_{D1} \approx v_{D2}$ Fewer non-deg. v-mismatched pairs

$\rightarrow$ Good $M_{157}{:}M_{268}$ reflection $\therefore$ Lower E_V & A_C

Balanced N-type $M_3{:}M_4$ mirror:

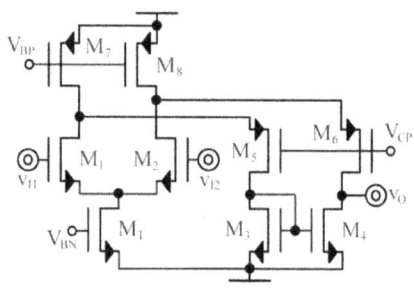

$\therefore$ $A_{DD} \approx 0$ V/V $A_{SS} \approx 1$ V/V

More non-degenerated pairs than in diff. stage:

$\therefore$ Higher v_{in}^*

$$v_{in}^* \approx \sqrt{2\left(\frac{i_{12}^*}{g_{m12}}\right)^2 + 2\left(\frac{i_{78}^*}{g_{m12}}\right)^2 + 2\left(\frac{i_{34}^*}{g_{m12}}\right)^2 + 2\left\{\left(\frac{i_{56}^*}{g_{m56}}\right)\left[\frac{g_{m56}}{1 + g_{m56}\left(r_{ds78} \parallel r_{ds12}\right)}\right]\left(\frac{1}{g_{m12}}\right)\right\}^2}$$

$\therefore$ Higher V_{OS}^*

$$V_{OS}^* \approx \sqrt{\left(\frac{\Delta i_{12}^*}{g_{m12}}\right)^2 + \left(\frac{\Delta i_{78}^*}{g_{m12}}\right)^2 + \left(\frac{\Delta i_{34}^*}{g_{m12}}\right)^2 + \left\{\left(\frac{\Delta i_{56}^*}{g_{m56}}\right)\left[\frac{g_{m56}}{1 + g_{m56}\left(r_{ds78} \parallel r_{ds12}\right)}\right]\left(\frac{1}{g_{m12}}\right)\right\}^2}$$

i. Reverse Fold

Concept: M_5 diode/mirror-connects M_3:M_4 $\rightarrow$ $-$ FB loop

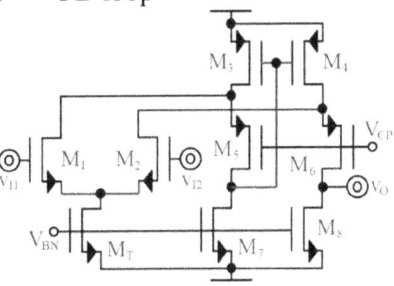

Bias: $I_3 = I_1 + I_7 \approx 0.5I_T + I_7$

$I_4 = I_2 + I_8 \approx 0.5I_T + I_8 \approx I_3$

Small v_{ID}: v_{S5} T-mixes $i_i = i_1$ & $i_{fb} = i_3$

M_5 current-buffers $i_e = i_3 - i_1 \approx 0$ $\qquad$ Advantage:

M_4 mirrors $i_3 = i_{fb} \approx i_1$ $\quad$ Diff. Conv. at v_{S6} $\qquad$ N pair feeds P mirror,

M_6 current-buffers $i_o = i_4 - i_2 \approx i_1 - i_2$ into v_O $\qquad$ like differential stage,

$\therefore$ $A_D = G_D R_O \approx g_{m12}(R_{D6} \parallel r_{ds8}) \approx g_{m12}r_{ds8}$ $\qquad$ with higher $v_{IC(MAX)}$.

C. Bias

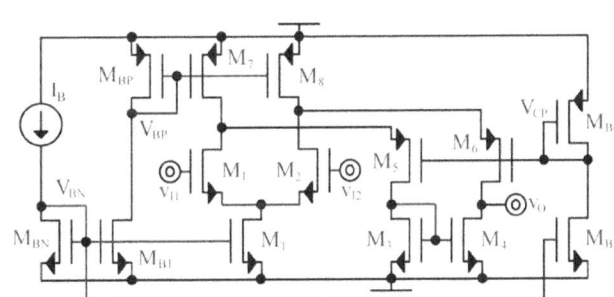

Mirror Translations:

$M_{BN} : M_{B1} : M_T : M_{B2}$

$M_{BP} : M_7 : M_8$

Requirement: $i_{78} > i_{12}$ $\qquad$ Design: $\quad$ If $i_{78} > i_T$ $\therefore$ $i_{3456} > 0$ during SR

Cascodes: $\qquad\qquad\qquad\qquad\qquad\qquad$ $\rightarrow$ M_{3456} do not cut off

Requirement: $v_{SD78} \geq v_{SD78(SAT)}$ $\qquad\qquad$ $\rightarrow$ Faster recovery

$\therefore$ $v_{SD78} = v_{SGBC} - v_{SG56} = v_{SDBC(SAT)} - v_{SD56(SAT)} + |V_{TP0}| - |V_{TP56}| \geq v_{SD78(SAT)}$
$\qquad\qquad\qquad\qquad\qquad\qquad\qquad\qquad\qquad\qquad\qquad\qquad\qquad\downarrow$

$\rightarrow$ V_{S56} limits $v_{IC(MAX)}$ & $v_{O(MAX)}$ $\qquad$ Body effect if $v_{SB56} \neq 0$

D. Folded Cascode

Modification: Cascode load mirror $M_3{:}M_4$

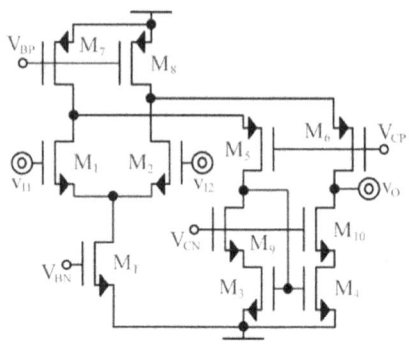

Advantages: Higher R_O

∴   Higher A_D

$$A_D = G_D R_O = g_{m12}(R_{D6} \| R_{D10})$$

$$\approx g_{m12}g_{m6}r_{ds6}(r_{ds2} \| r_{ds8})$$

Low v_D mismatch across mirror   →   Lower $V_{OS(S)}$

→   Lower A_C $f(V_{CN})$
                      ↑

Variants: Low-voltage cascode mirror   → $v_O \geq v_{DS10(SAT)} + v_{DS4} + v_{SS}$

Basic cascode mirror   → $v_O \geq v_{DS10(SAT)} + v_{GS34} + v_{SS}$

Variations: N/P pair, mirror/cascode fold, N/P basic/cascode (basic/low-v) mirror

Example

Given: $C_{LD} = 1$ pF, $v_{DD} = 2$ V, $i_{78} = i_T = 60$ μA, $V_{SD34} = V_{DS78} = 400$ mV, L's = 1 μm,

$W_{N/P}$'s = 5/15 μm, $K_{N/P}' = 200/40$ μA/V^2, $V_{TN0} = -V_{TP0} = 400$ mV, $n_I = 1.75$,

$1/\lambda_N = 1/\lambda_P = 50$ V when L = 1 μm, $t_{OX} = 5$ nm, $L_{OL} = 30$ nm.

Bias: $L_{CH} = L - 2L_{OL} = 940$ nm $I_{349/10} = I_{78} - I_{12} \approx I_{78} - 0.5i_T = 30$ μA

$V_{SDT(SAT)} = 430$ mV

$V_{SD12(SAT)} = 310$ mV

$V_{DS78(SAT)} = 340$ mV

$V_{DS56(SAT)} = 240$ mV

$V_{SD349/10(SAT)} = 310$ mV

$> 2n_I V_t$

$= 90$ mV

∴   Strong

Inversion

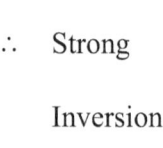
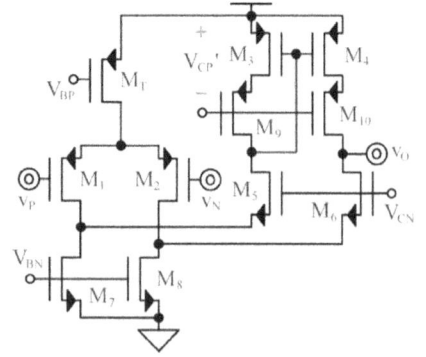

$V_{CN} = V_{DS78} + V_{TN0} + V_{DS56(SAT)} = 1.04$ V $V_{CP}' = V_{SD34} + |V_{TP0}| + V_{SD9/10(SAT)} = 1.11$ V

Large Signal:

$$v_{IC} \leq v_{DD} - V_{SDT(SAT)} - |V_{TP0}| - V_{SD12(SAT)} = 860 \text{ mV}$$

$$v_{IC} \geq V_{DS78} - |V_{TP0}| = 0.0 \text{ V}$$

$$v_O \leq v_{DD} - V_{SD4} - V_{SD10(SAT)} = 1.29 \text{ V}$$

$$v_O \geq V_{DS78} + V_{DS6(SAT)} = 640 \text{ mV}$$

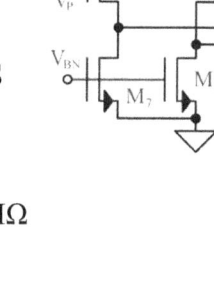

Small Signal:

$$g_{m12} = 200 \text{ μS} \qquad g_{m349/10} = 200 \text{ μS}$$

$$g_{m56} = 250 \text{ μS} \qquad r_{ds12} = 1.7 \text{ MΩ}$$

$$r_{ds78} = 830 \text{ kΩ} \qquad r_{ds34569/10} = 1.7 \text{ MΩ}$$

$$C_{GD12349/10T} = 3.1 \text{ fF} \qquad C_{GD5678} = 1.0 \text{ fF}$$

$$C_{GS12349/10T} = 68 \text{ fF} \qquad C_{GS5678} = 23 \text{ fF}$$

$$R_O = R_{D10} \| R_{D6} = 580M \| 360M = 220 \text{ MΩ}$$

$$A_D = g_{m12}R_O = 44 \text{ kV/V} = 93 \text{ dB}$$

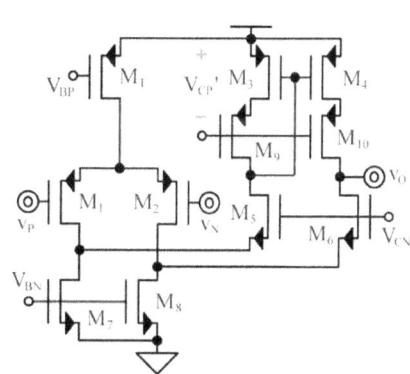

$$p_O \approx \frac{1}{2\pi R_O \left(C_{GD10} + C_{GD6} + C_{LD} \right)} = 720 \text{ Hz}$$

$$f_{0dB} \approx GBW = A_D p_O = 32 \text{ MHz}$$

$$p_M \approx \frac{g_{m34}}{2\pi \left(C_{GS3} + C_{GD3} + C_{GS4} + C_{GD4} + C_{GD5} \right)} = 220 \text{ MHz}$$

$$p_F \approx \frac{g_{m56}}{2\pi \left(C_{GD12} + C_{GD78} + C_{GS56} \right)} = 1.5 \text{ GHz} \approx f_{T56} \qquad z_{MX} \approx 2p_M \approx 440 \text{ MHz}$$

$$PM = 180° - \tan^{-1}\left(\frac{f_{0dB}}{p_O} \right) - \tan^{-1}\left(\frac{f_{0dB}}{p_M} \right) + \tan^{-1}\left(\frac{f_{0dB}}{z_M} \right) - \tan^{-1}\left(\frac{f_{0dB}}{p_F} \right) = 85°$$

Simulation:

* P-Input Folded Cascode

mbn vbn vbn vss vss nmos1 w=5u l=1u

mb1 vbp vbn vss vss nmos1 w=5u l=1u

mbp vbp vbp vdd vdd pmos1 w=15u l=1u

mt vs vbp vdd vdd pmos1 w=15u l=1u

m1 v1 vp vs vs pmos1 w=15u l=1u

m2 v2 vn vs vs pmos1 w=15u l=1u

m7 v1 vbn vss vss nmos1 w=5u l=1u

m8 v2 vbn vss vss nmos1 w=5u l=1u

m5 vg vcn v1 v1 nmos1 w=5u l=1u

m6 vo vcn v2 v2 nmos1 w=5u l=1u

m9 vg vcp v3 v3 pmos1 w=15u l=1u

m10 vo vcp v4 v4 pmos1 w=15u l=1u

m3 v3 vg vdd vdd pmos1 w=15u l=1u

m4 v4 vg vdd vdd pmos1 w=15u l=1u

vcn vcn vss dc=1.05

vcp vdd vcp dc=1.12

ib vdd vbn dc=60u

cld vo 0 1p

vdd vdd 0 dc=2 ac=0

vss vss 0 dc=0 ac=0

efb vox vss vo vss 1

r2 vox vfb 1e3

r1 vfb vss 1e3

rb vfb vn 1m

vin vp vss dc=600m ac=1 sin 600m 50m 1e6

.model nmos1 nmos vto=0.4 kp=200u lambda=20m
+ tox=5n ld=30n cgso=207p cgdo=207p

.model pmos1 pmos vto=-0.4 kp=40u lambda=20m
+ tox=5n ld=30n cgso=207p cgdo=207p

Test Circuit

$0 \leq v_{IC} \leq 830$ mV

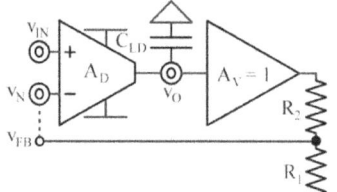

650 mV $\leq v_O \leq 1.3$ V

$V_O \equiv 2V_{FB} \approx V_G \approx 1.2$ V $< v_{O(MAX)}$

R_{12} reduces A_D ∴ Buffer v_O

.op

*.dc vin 0 2 1m

*.ac dec 100 1 10e9

*lb vfb vn 10e6

*cd vn vss 10e6

*cc vn vp 10e6

*.tran 5u

.end

5.3. Output: A. PSR Voltage-Divider Model

Model what connects to v_O:

$i_{gt/b}$'s inject or remove $v_{dd/ss}$

$R_{T/B}$'s couple (voltage-divide) $v_{dd/ss}$

$C_{T/B}$'s couple $v_{dd/ss}$

Z_{LD} shunts $v_{dd/ss}$

Biasing Mirrors:

If R_B = Very high $\quad \therefore \quad i_{gO} \approx \dfrac{v_{dd \cdot ss}}{R_B + 1/g_{mB}} \quad \rightarrow \quad$ Very low

CC/D Followers:

Reproduce $v_{b/g}$ $\therefore$ Remove $v_{b/g}$ noise

CE/S Transconductors: $\quad$ Common Mode

Amplify $v_{be/gs}$ $\quad \therefore$ $v_{b/g}$ should reproduce $v_{e/s}$ noise

B. Class A

Consider: $\quad$ Symmetrical Supplies $\rightarrow$ $v_{DD} = |v_{SS}|$

$\quad\quad\quad\quad$ Sinusoidal v_O

Class-A Transistor:

$\quad\quad\quad$ Conducts 360° of sine wave $\quad \rightarrow \quad$ Conduction Angle = 360°

Low $P_A = v_{DS}i_A$ with: $\quad$ Low v_{DS} $\quad\quad \rightarrow \quad\quad$ High $v_{O(PK)}$

$\quad\quad\quad\quad\quad\quad\quad\quad\quad\quad\quad$ Low i_A $\quad\quad\quad \rightarrow \quad\quad i_{A(MIN)} \approx 0$ when v_{DS} maxes

Low-Side M_A: $\quad\quad\quad i_{A(MIN)} = 0$ when $i_{O(MAX)} \equiv i_{O(PK)} \equiv I_Q$

$\therefore \quad i_{O(MIN)} = -i_{O(MAX)} = -I_Q \quad\quad i_{A(MAX)} = 2I_Q \quad\quad v_{O(PK)} = i_{O(PK)}R_{LD} = I_Q R_{LD}$

Power Efficiency: $\quad\quad\quad\quad\quad\quad\quad\quad\quad\quad\quad\quad\quad\quad \leq 1 \quad\quad \eta_{A(OPT)}$

$$\eta_A = \frac{P_O}{P_{SUPPLIES}} = \frac{v_{O(RMS)}i_{O(RMS)}}{(v_{DD} - v_{SS})i_{AVG}} = \frac{\left(\dfrac{v_{O(PK)}}{\sqrt{2}}\right)\left(\dfrac{i_{O(PK)}}{\sqrt{2}}\right)}{2v_{DD}I_Q} = \frac{1}{4}\left(\frac{v_{O(PK)}}{v_{DD}}\right)\left(\frac{i_{O(PK)}}{I_Q}\right) \leq 25\%\left(\frac{v_{O(PK)}}{v_{DD}}\right)$$

$$\eta_{A(MAX)}$$

i. CC/D Follower

Power Efficiency:

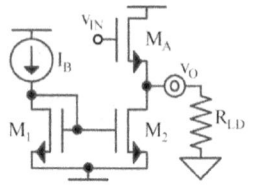

$$v_{O(PK)} = v_{O(MAX)} = v_{IN(MAX)} - v_{GSA}$$

$$\leq (v_{DD} - V_{DS(SAT)}) - (v_{TNA} + V_{DSA(SAT)}) < v_{DD}$$

Body effect increases v_{TNA} → Reduces $v_{O(MAX)}$

$$\therefore \quad \eta_{A(CC/D)} < \eta_{A(MAX)} = 25\%$$

Power-Supply Rejection: v_{IN} carries v_{dd} or v_{ss} (from differential stage)

$\therefore$ i_{gA} reproduces v_{dd} or v_{ss} $\qquad\qquad$ i_{g2} mirrors v_{dd} & v_{ss} that I_B's R_B injects

$\quad$ r_{dsA} & r_{ds2} couple v_{dd} & v_{ss} $\qquad\qquad$ $1/g_{mA}$ & R_{LD} shunt r_{dsA}, r_{ds2}, & i_{g2} noise

$$v_o \approx \frac{v_{dd/ss}\,g_{mA}\left(r_{ds2} \| R_{LD}\right)}{1+\left(g_{mA}+g_{mbA}+1/r_{dsA}\right)\left(r_{ds2} \| R_{LD}\right)} \qquad\qquad \text{CC/D reproduces } v_g \text{ noise}$$

$$+\left(\frac{v_{dd}}{r_{dsA}+\left(1/g_{mA} \| R_{LD}\right)}+\frac{v_{ss}}{r_{ds2}+\left(1/g_{mA} \| R_{LD}\right)}-\frac{v_{dd}}{R_B}+\frac{v_{ss}}{R_B}\right)\left(\frac{1}{g_{mA}} \| R_{LD}\right) \approx v_{dd/ss}$$

Example: Design W_2 & W_A so $\eta_{A(CD)} \equiv 18\%$ when $L_{MIN} = 180$ nm, $L_{OL} = 30$ nm,

$v_{IN(MAX)} \approx v_{DD} = -v_{SS} = 2.5$ V, $v_{BS} = 0$, $V_{TN0} = 400$ mV, $K_N' = 200$ μA/V², $R_{LD} = 1$ kΩ.

Solution:

$$\eta_{A(CD)} = 25\%\left(\frac{v_{O(PK)}}{v_{DD}}\right)\left(\frac{i_{O(PK)}}{I_Q}\right) \equiv 18\% \qquad \rightarrow \qquad I_Q \equiv i_{O(PK)} \qquad \therefore \qquad v_{O(PK)} = 1.8 \text{ V}$$

$$I_2 = I_Q \equiv i_{O(PK)} = \frac{v_{O(PK)}}{R_{LD}} = 1.8 \text{ mA} \quad \rightarrow \quad i_{A(MAX)} = 2I_Q = 3.6 \text{ mA}$$

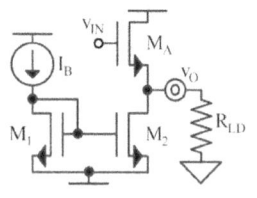

$$v_{O(MIN)} = v_{SS} + V_{DS2(SAT)} \equiv -v_{O(PK)} = -1.8 \text{ V}$$

$$\rightarrow \quad V_{DS2(SAT)} \approx \sqrt{\frac{2I_2}{K_N'(W_2/L_2)}} \leq 700 \text{ mV} \quad \therefore \quad L_2 + 2L_{OL} \equiv 5L_{MIN} = 900 \text{ nm}$$

$$\uparrow$$

$$\text{Low } \lambda \text{ error} \qquad W_2 \geq 31 \text{ μm}$$

$$v_{O(MAX)} = v_{IN(MAX)} - v_{TN} - V_{DSA(SAT)} \equiv 1.8 \text{ V}$$

$$v_{DD} \approx \hookleftarrow \qquad\qquad\qquad\qquad\qquad\qquad \text{High } f_{TA} \qquad W_A \geq 48 \text{ μm}$$

$$\downarrow$$

$$\rightarrow \quad V_{DSA(SAT)} \approx \sqrt{\frac{2i_{A(MAX)}}{K_N'(W_A/L_A)}} \leq 300 \text{ mV} \quad \therefore \quad L_A + 2L_{OL} \equiv L_{MIN} = 180 \text{ nm}$$

ii. CE/S Transconductor

Power Efficiency:

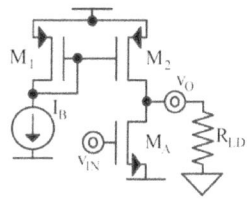

$$v_{O(PK)} = |v_{O(MIN)}| = |v_{SS} + V_{DSA(SAT)}| < |v_{DD/SS}|$$

$$\therefore \quad \eta_{A(CE/S)} < \eta_{A(MAX)} = 25\%$$

No body effect

Power-Supply Rejection: M_A's source receives v_{ss} $\therefore$ v_{IN} should carry v_{ss}

$\therefore$ v_{gsA} cancels v_{ss} $\qquad\qquad$ r_{ds2} & r_{dsA} couple v_{dd} & v_{ss}

$\rightarrow$ $i_{gA} \approx 0$ $\qquad\qquad\qquad$ i_{g2} mirrors v_{dd} & v_{ss} that I_B's R_B injects

R_{LD} shunts/reduces v_O noise $\qquad\qquad\qquad\qquad\qquad$ Usually
$$\downarrow$$

$$v_o \approx \frac{v_{dd}\left(r_{dsA} \| R_{LD}\right)}{r_{ds2} + \left(r_{dsA} \| R_{LD}\right)} + \frac{v_{ss}\left(r_{ds2} \| R_{LD}\right)}{r_{dsA} + \left(r_{ds2} \| R_{LD}\right)} + \left(\frac{v_{dd}}{R_B} - \frac{v_{ss}}{R_B}\right)(r_{dsA} \| r_{ds2} \| R_{LD}) < v_{dd/ss}$$

Example: Determine $\eta_{A(CS)}$ & W_2 when $W_A = 48\ \mu m$, $L_A = L_{MIN} - 2L_{OL} = 120$ nm,

$V_{TN0} = 400$ mV, $K_N' = 200\ \mu A/V^2$, $K_P' = 40\ \mu A/V^2$, $v_{DD} = -v_{SS} = 2.5$ V, $R_{LD} = 1\ k\Omega$.

Solution: $\quad I_Q \equiv i_{O(PK)}$ $\quad \therefore \quad i_{A(MIN)} = 0$ $\qquad i_{A(MAX)} = 2I_Q$ $\qquad v_{O(PK)} = I_Q R_{LD}$

$$v_{O(MIN)} = v_{SS} + V_{DSA(SAT)} \approx v_{SS} + \sqrt{\frac{2\left(2I_Q\right)}{K_N'(W_A/L_A)}} \equiv -v_{O(PK)} = -I_Q R_{LD}$$

$\therefore$ $I_Q = 2.15$ mA $\qquad i_{A(MAX)} = 4.3$ mA $\qquad v_{O(PK)} = 2.15$ V

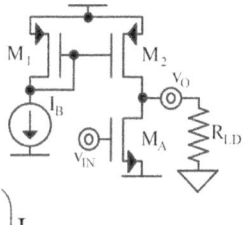

$$\therefore \quad \eta_{A(CS)} = 25\%\left(\frac{v_{O(PK)}}{v_{DD}}\right)\left(\frac{i_{O(PK)}}{I_Q}\right) = 21.5\%$$

$$v_{O(MAX)} = v_{DD} - V_{SD2(SAT)} \equiv v_{O(PK)} = 2.15\ V$$

$$W_1 = \left(\frac{W_2}{L_2}\right)\left(\frac{I_B}{I_2}\right)L_1$$

$$W_2 \geq 790\ \mu m$$

Low λ error
$$\downarrow$$

$$V_{SD2(SAT)} \approx \sqrt{\frac{2I_2}{K_P'(W_2/L_2)}} \leq 350\ mV \quad \therefore \quad L_{12} + 2L_{OL} \equiv 5L_{MIN} = 900\ nm$$

Distortion

Definition: Effects on pure sinusoid $\rightarrow$ $v_{in} = V_P\sin(\omega_O t)$

$$2\pi f_O$$

Distorted: $v_o = A_1 V_P\sin(\omega_O t) + A_2 V_P\sin(2\omega_O t) + \dots + A_N V_P\sin(N\omega_O t)$

Harmonic Distortion:

$$HD_N \equiv \left|\frac{A_N}{A_1}\right|$$

Harmonic-to-fundamental gain ratio

Total Harmonic Distortion:

$$THD \equiv \sqrt{\left(\frac{A_2}{A_1}\right)^2 + \left(\frac{A_3}{A_1}\right)^2 + \dots + \left(\frac{A_N}{A_1}\right)^2}$$

Root–sum–squared (RSS)

CC/D Voltage Followers: $\Delta v_O \approx \Delta v_{IN}$ $\rightarrow$ Largely linear $\therefore$ Low gain distortion

CE/S Transconductors:

Small signals $\approx$ Linear v_{in} translations $\therefore$ Low gain distortion

Large signals = Exponential or quadratic translations $\therefore$ Distorted

C. Class B/AB

Class-B Transistor: Conducts 180° of sinusoid

$\therefore$ $i_B = i_{B(MIN)} = 0$ across + or – half cycle

$i_{B(MAX)} = i_{O(PK)} = v_{O(PK)}/R_{LD}$ when $v_O = v_{O(PK)}$

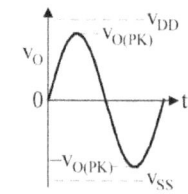

Class-AB Transistor: Conducts 180° < Angle < 360° of sinusoid

$i_{AB} > 0$ when $v_O = 0$ (across 0-V half-cycle crossings)

Class-B/AB Stage: Push–pull B/AB transistors

M_T supplies P_{VDD} across + half cycles

M_B supplies P_{VSS} across – half cycles

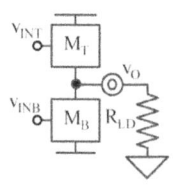

Power Efficiency:

$$\eta_B = \frac{v_{O(RMS)}i_{O(RMS)}}{v_{DD}i_{DD(AVG)} + |v_{SS}|i_{SS(AVG)}} = \frac{\left(\dfrac{v_{O(PK)}}{\sqrt{2}}\right)\left(\dfrac{i_{O(PK)}}{\sqrt{2}}\right)}{2v_{DD}\left(\dfrac{i_{O(PK)}}{\pi}\right)} = \frac{\pi}{4}\left(\frac{v_{O(PK)}}{v_{DD}}\right) = \overbrace{78.5\%\left(\frac{v_{O(PK)}}{v_{DD}}\right)}^{\eta_{B(MAX)}}$$

$\llcorner$ ½ Sine $\lrcorner$

i. CC/D Followers

η_B: $\qquad$ $v_{BE/GS}$'s limit $v_{O(PK)}$

PSR: $\qquad$ M_T & M_B reproduce v_{dd} or v_{ss} in v_{IN}

Distortion: $\qquad$ Followers $\approx$ Linear when $i_{C/D}$'s > 0

$\qquad\qquad$ $i_{C/D}$'s rise with higher $V_{BAT} = v_{GT} - v_{GB} = V_{BN} + V_{BP}$

$\therefore$ $\qquad$ Class B with low $v_{GT} - v_{GB}$ $\qquad\qquad$ Class AB with high $v_{GT} - v_{GB}$

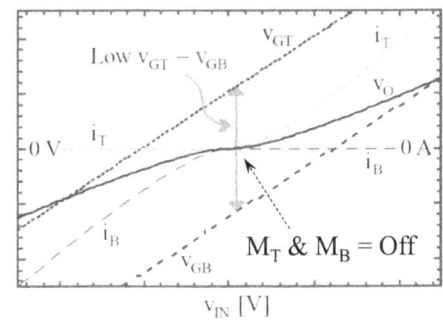 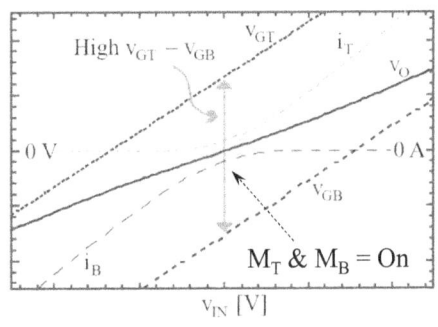

Each transistor conducts $\approx 180°$ $\qquad$ Each transistor conducts > $180°$

Cross-over distortion when i_D's = 0 $\qquad$ i_D's > 0 when $v_O = 0$ → Lower X distortion

Example: Diode Stack ("Diamond" Driver)

Architecture:

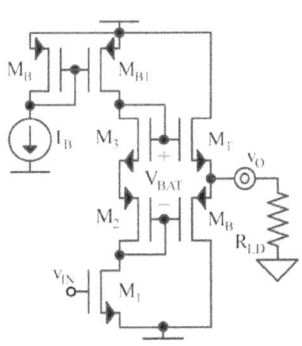

Push–Pull Output → M_T : M_B

Bias Current → I_B : M_B : M_{B1}

Class-B/AB Battery → M_2 : M_3 Diode Stack

CS Amplifying Driver → M_1

I_{B1} & v_{GS} loop set i_T & i_B:

$\qquad$ $v_{SG2} + v_{GS3} = \text{Constant} \equiv V_{BAT} = f(I_{B1})$ $\therefore$

$\qquad$ $V_{SD2(SAT)} + V_{DS3(SAT)} = v_{SDB(SAT)} + v_{DST(SAT)}$

Operation: When M_B pulls i_{LD}: $\qquad$ $v_{SGB} = \text{Higher}$ $\therefore$ V_{BAT} reduces v_{GST} to set i_T

$\qquad\qquad$ When M_T supplies i_{LD}: $v_{GST} = \text{Higher}$ $\therefore$ V_{BAT} reduces v_{SGB} to set i_B

$\qquad\qquad$ When $v_O = 0$ V: $i_{LD} = 0$ $\therefore$ $i_T = i_B$ → V_{BAT} sets v_{GST} & v_{SGB} to set i_{TB}

$\qquad\qquad\qquad$ If $V_{BAT} = \text{Low}$ $\therefore$ $i_{TB} = 0$ when $v_O = 0$ → Class B

$\qquad\qquad\qquad$ If $V_{BAT} = \text{High}$ $\therefore$ $i_{TB} > 0$ when $v_O = 0$ → Class AB

ii. CE/S Transconductors

η_B: $V_{DS(SAT)}$'s limit $v_{O(PK)}$

PSR: M_T or M_B cancels & other amplifies v_{dd} or v_{ss} in v_{IN}

Distortion: $i_{C/D} \propto \exp v_{IN}$ or v_{IN}^2 $\rightarrow$ Gain distortion

$\qquad\qquad$ $i_{C/D}$'s fall with higher $V_{BAT} = v_{GT} - v_{GB} = V_{BN} + V_{BP}$

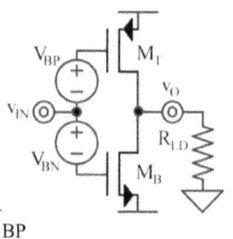

$\therefore$ Class B with high $v_{GT} - v_{GB}$ $\qquad\qquad$ Class AB with low $v_{GT} - v_{GB}$

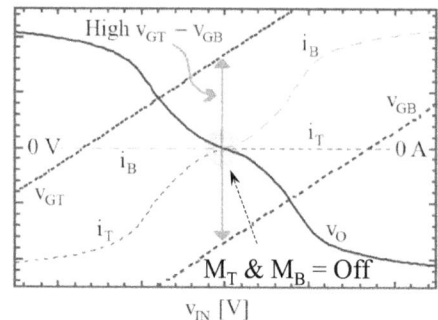 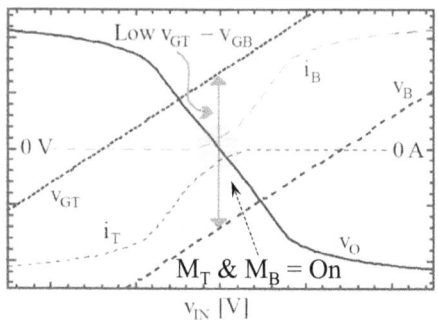

Each transistor conducts $\approx 180°$ $\qquad\qquad$ Each transistor conducts $> 180°$

Cross-over distortion when i_D's $= 0$ $\qquad$ i_D's > 0 when $v_O = 0 \rightarrow$ Lower X distortion

Example: Adaptive Stack
Operation

$\qquad$ When M_B pulls i_{LD}: $v_{GSB} =$ High

$\qquad\quad \therefore$ M_3 shuts & M_4 current-buffers i_2

$\qquad\qquad$ V_{GP} & v_{SG4} with I_2 bias M_T to $i_{T(MIN)}$

$\qquad$ When M_T sources i_{LD}: $v_{SGT} =$ High

$\qquad\quad \therefore$ M_4 shuts & M_3 current-buffers i_1

$\qquad\qquad$ V_{GN} & v_{GS3} with I_1 bias M_B to $i_{B(MIN)}$

Push–Pull Output $\rightarrow$ $M_T : M_B$

B/AB Bias $\rightarrow$ $M_3 : M_4$

CS amplifying drivers $\rightarrow$ $M_1 : M_2$

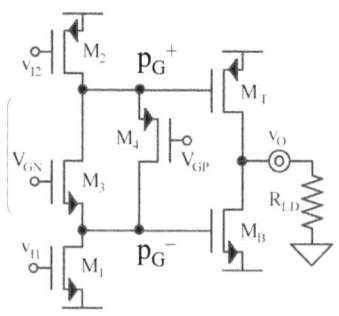

Bias: V_{GN} & M_3 bias M_B $\qquad$ Headroom $\equiv$ Min$\{v_{DD} - v_{SS}\} \geq 2|v_{GS}| + |V_{DS(SAT)}|$

$\qquad$ V_{GP} & M_4 bias M_T $\qquad$ p_G at gates: M_{34} close + FB loop $\rightarrow$ High R_G, Low f_O

Class AB:	$V_{GN} > v_{SS} + 2v_{TN} + V_{DS3(SAT)}$	Class B:	$V_{GN} \leq \ldots$
$i_{T/B(MIN)} > 0$	$V_{GP} < v_{DD} - 2|v_{TP}| - V_{SD4(SAT)}$	$i_{T/B(MIN)} = 0$	$V_{GP} \geq \ldots$

78

D. Summary

Conduction Angle: $\qquad \angle A \qquad > \qquad \angle AB \qquad > \qquad \angle B$

Higher Power Efficiency:

With lower conduction $\qquad \therefore \qquad \eta_B \qquad > \qquad \eta_{AB} \qquad > \qquad \eta_A$

With lower $v_{DS/CE} \quad \rightarrow \quad$ Higher $v_{O(PK)}$

$$\eta_B = \frac{\pi}{4}\left(\frac{v_{O(PK)}}{v_{DD}}\right) > \eta_A = \frac{1}{4}\left(\frac{v_{O(PK)}}{v_{DD}}\right)\left(\frac{i_{O(PK)}}{I_Q}\right) \equiv \frac{1}{4}\left(\frac{v_{O(PK)}}{v_{DD}}\right) \qquad \begin{array}{l}\text{When } I_Q \equiv i_{O(PK)} \\[4pt] \therefore \qquad i_{A(MIN)} = 0 \end{array}$$

$$\therefore \quad \eta_{\text{Transconductor}} \quad > \quad \eta_{\text{Follower}} \qquad\qquad\qquad\qquad i_{A(MAX)} = 2I_Q$$

Distortion:

Nonlinear v_{IN} translations distort signals $\qquad \rightarrow \qquad$ Gain distortion

$\therefore \quad THD_{\text{Transconductor}} \qquad > \qquad THD_{\text{Follower}}$

$\qquad THD_{\text{Large Signals}} \qquad > \qquad THD_{\text{Small Signals}} \qquad \rightarrow \qquad$ Better with lower Δv_O

Cross-over distortion worsens with lower conduction $\qquad\qquad$ in output stage

$\therefore \quad THD_B \qquad > \qquad THD_{AB} \qquad > \qquad THD_A$

5.4. Class-A (Classic) Op-Amp Example: A. Static

Bias: $\qquad$ – FB mixes v_P & v_N $\qquad \therefore \qquad v_N \approx v_P \qquad v_O \approx \dfrac{v_P}{\beta_{FB}}$

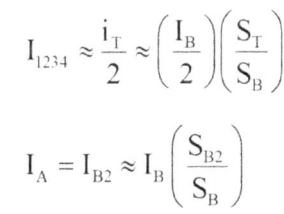

$$I_{1234} \approx \frac{i_T}{2} \approx \left(\frac{I_B}{2}\right)\left(\frac{S_T}{S_B}\right)$$

$$I_A = I_{B2} \approx I_B\left(\frac{S_{B2}}{S_B}\right)$$

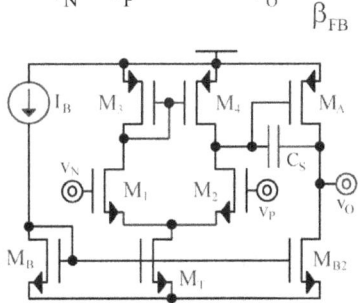

v_{IC} Range: $\quad$ Same as differential stage

v_O Range: $\quad v_{SS} + v_{DSB2(SAT)} \leq v_O \leq v_{DD} - v_{SDA(SAT)} \quad \rightarrow \quad$ Rail-to-Rail

i_O Range: $\quad -I_{B2} < i_O < i_{A(MAX)} - I_{B2} \qquad\qquad (v_{IC} - v_{GS2}) + v_{DS2(SAT)}$

$\qquad\qquad\qquad\qquad\qquad\qquad\qquad\qquad\qquad\qquad\qquad \uparrow$

$i_{A(MAX)} = f(\text{Gate Drive}) = f(v_{SGA(MAX)}) = f(v_{DD} - v_{GA(MIN)}) = f(v_{DD} - (v_{IC} - v_{TN2}))$

Stabilization: $\quad C_S$ splits $p_1 = p_{GA} : p_2 = p_O \qquad \therefore \qquad v_{IC(MAX)}$ limits $i_{O(MAX)}$

V_{OS}: If $V_{SG3} \approx V_{SGA}$ $\therefore$ $V_{D13} \approx V_{D24}$

$$V_{OS(S)} = \frac{\Delta v_D}{A_{V10}} = \frac{v_{D13} - v_{D24}}{A_{V10}} = \frac{V_{SG3} - V_{SGA} \pm (\Delta v_O / A_{V20})}{A_{V10}} \approx \frac{\pm \Delta v_O}{A_{V10} A_{V20}} \rightarrow \text{Very Low}$$

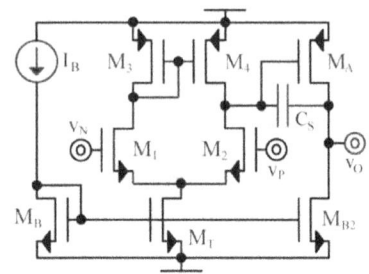

$\therefore$ Design Aim: $V_{SD3(SAT)} \equiv V_{SDA(SAT)}$

$$\sqrt{\frac{2I_3}{K_P'S_3}} \equiv \sqrt{\frac{2I_A}{K_P'S_A}}$$

$$\frac{I_3}{S_3} \equiv \frac{I_A}{S_A} \rightarrow \text{Current densities match}$$

$$V_{OS}^* \approx \sqrt{\left(\frac{\Delta i_{12}^*}{g_{m12}}\right)^2 + \left(\frac{\Delta i_{34}^*}{g_{m12}}\right)^2 + \left(\frac{\Delta i_{A(3A)}^*}{g_{mA}A_{V10}}\right)^2 + \left(\frac{\Delta i_{B2(TB2)}^*}{g_{mA}A_{V10}}\right)^2} \approx \sqrt{\left(\frac{\Delta i_{12}^*}{g_{m12}}\right)^2 + \left(\frac{\Delta i_{34}^*}{g_{m12}}\right)^2}$$

B. Dynamic: i. Large Signal

SR: Largest C's limit SR $\rightarrow$ Intentional C_S & C_{LD}

$$SR_S \propto \Delta(v_O - v_{GA}) = \Delta v_O (1 - 1/A_{VA}) \approx \Delta v_O \quad \therefore \quad SR_S \approx SR_{LD} = SR_O$$

Scenarios: C_S by i_2 or i_4 $\rightarrow$ $SR_S^\pm = i_T/C_S$ Design Aim:

C_{LD}^+ by $i_A^+ - i_2 - I_{B2}$ $\rightarrow$ $SR_{LD}^+ = (i_A^+ - i_T - I_{B2})/C_{LD}$ $\rceil$ $i_A^+ \gg i_T + I_{B2}$

C_{LD}^- by $I_{B2} - i_4$ $\rightarrow$ $SR_{LD}^- = (I_{B2} - i_T)/C_{LD}$ $\rfloor$ $I_{B2} \gg i_T$

SR$^+$:   SR$^-$:

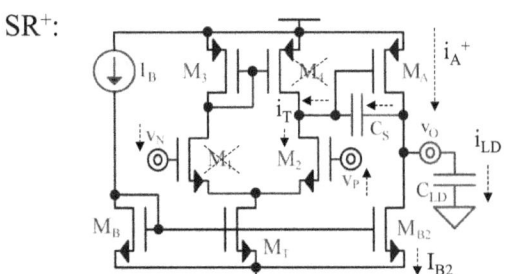
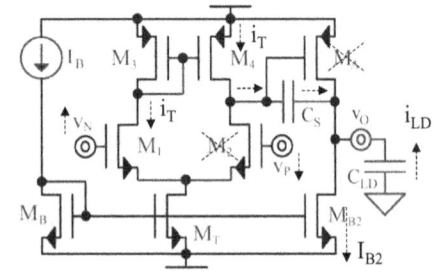

Worst-Case $SR_O^\pm = \text{Min}\{SR^{\pm}\text{'s}\} \leq i_T/C_S$

ii. Small Differential Signals

$R_{ID} \to \infty$ $R_{O1} = r_{ds2} \parallel r_{ds4}$

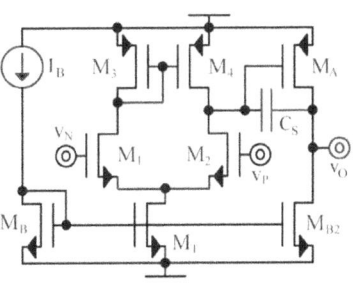

$R_O = r_{dsA} \parallel r_{dsB2}$ $A_{V0} = g_{m12} R_{O1} g_{mA} R_O$

Split M_A's C_X: $C_{XI} : C_{XO}$ $A_{VA0} = -g_{mA} R_O$

p_1 OC Model: $p_1 \approx p_S \approx \dfrac{1}{2\pi R_{O1}(1 - A_{VA0}) C_S} \approx \dfrac{1}{2\pi R_{O1} g_{mA} R_O C_S}$ $\to$ Low

$$\text{If } p_S \ll f_{0dB} \le p_O, z_S, p_M, z_{MX} \quad \therefore \quad f_{0dB} \approx GBW = A_{V0} p_S \approx \frac{g_{m12}}{2\pi C_S}$$

p_2 SC Model: C_{GA} shunts R_{O1} past p_S $\to$ R_{O1} fades

FB Effect: No i_{FW} $\to$ No v_{ID}

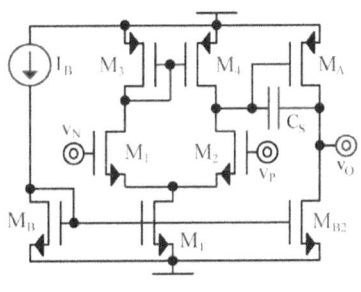

C_S & M_A close – FB loop

C_S voltage-divides v_o into C_{GA}

$\to$ C_S diode-connects M_A

$$\therefore \quad p_O = \frac{1}{2\pi R_{OA} C_O} \approx \left(\frac{g_{mA} C_S}{C_S + C_{GA}} \right) \left\{ \frac{1}{2\pi \left[C_{LD} + \left(C_S \oplus C_{GA} \right) \right]} \right\} \approx \frac{g_{mA}}{2\pi C_{LD}}$$

FW Effect: $C_{XI} : i_{FW}$

$A_{GA} = -g_{mA} + s(C_S + C_{GDA})$ $\to$ Out-of-phase $z_S \approx \dfrac{g_{mA}}{2\pi C_S}$

Mirror: $p_M = p_{G3} \approx \dfrac{g_{m3}}{2\pi \left(2 C_{GS34} \right)}$ In-phase $z_{MX} \approx 2 p_M \approx f_{T34}$

C. Power-Supply Rejection

M_A: $M_3 : M_4$ reproduces v_{dd} → $v_{gA} \approx v_{dd}$

v_{SGA} cancels v_{dd} → $i_{gA} \approx 0$

r_{dsA} couples v_{dd} into r_{dsB2}

$$A_{DD0} \approx \frac{r_{dsB2}}{r_{dsA} + r_{dsB2}} \rightarrow \frac{r_{dsB2}}{1/g_{mA} + r_{dsB2}} \approx 1$$

M_{B2}: If I_B's $R_B \gg r_{dsA/B2}$ → $v_{gB} \approx v_{ss}$ → Low i_{gB2}

$\left.\begin{array}{l}\end{array}\right\}$ $f_O > p_1 \approx p_S$

r_{dsB2} couples v_{ss} into r_{dsA}

$$A_{SS0} \approx \frac{r_{dsA}}{r_{dsB2} + r_{dsA}} \rightarrow \frac{1/g_{mA}}{r_{dsB2} + 1/g_{mA}} \ll 1$$

C_{LD}: Shunts A_V & $A_{DD/SS}$

∴ No effect in $PSRR = \dfrac{A_V}{A_{DD/SS}}$

C_S: Diode-connects M_A ∴ $Z_A \rightarrow \dfrac{1}{g_{mA}}$

∴ $\left.\begin{array}{l}\text{Couples } v_{dd} \\[6pt] \text{Shunts } v_{ss}\end{array}\right\}$ $PSRR^- > PSRR^+$

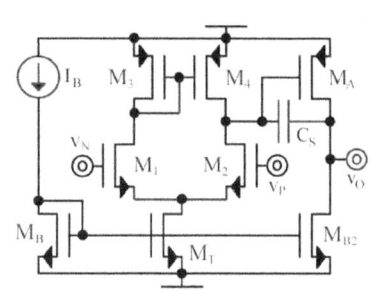

D. Nulling Zero

Requirement: $z_S \approx \dfrac{g_{mA}}{2\pi C_S} \gg f_{0dB} \approx \dfrac{g_{m12}}{2\pi C_S}$ → $g_{mA} \gg g_{m12}$ → High I_A

Alternative: Shift z_S to ∞ → Impede i_S Reverse p_S → Current-limit C_S

Design: $z_{SX} \approx \dfrac{1}{2\pi\left(R_S - 1/g_{mA}\right)C_S} \equiv p_O \approx \dfrac{g_{mA}}{2\pi C_{LD}}$

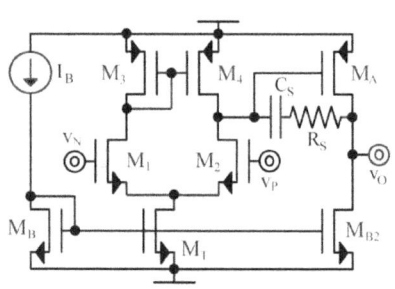

∴ $R_S \equiv \left(\dfrac{1}{g_{mA}}\right)\left(\dfrac{C_{LD} + C_S}{C_S}\right)$

SC: R_{O1} fades past p_S

C_S shorts & R_O fades past $z_{SX} \equiv p_O$

$$\frac{p_R}{f_{0dB}} \approx \frac{2\pi\left(1/g_{m12}\right)C_S}{2\pi R_S C_{GA}} \gg 1 \leftarrow \text{Usually}$$

∴ C_{GA} shunts R_S past $p_R = p_{GA}' \approx \dfrac{1}{2\pi R_S C_{GA}}$ → $p_S \ll f_{0dB}$, $p_O \approx z_{SX} < p_{M/R}$, z_{MX}

MOS Implementation

If moderate value, low-v_R coefficient, low-T_J drift R's $\neq$ available $\rightarrow$ Use R_{CH}

Invert channel $\rightarrow$ $v_{GSR} > v_{TN}$ or $v_{SGR} > |v_{TP}|$

Triode $\rightarrow$ M_R's $I_R = 0$ $\therefore$ $V_{SDR} = 0$

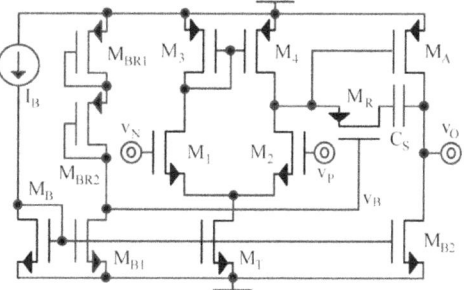

Design: $R_S = \dfrac{1}{K_P'(W_R/L_R)(v_{SGR} - |v_{TPR}|)}$

$\qquad = \dfrac{1}{K_P'(W_R/L_R)v_{SDR(SAT)}}$

$v_{SGBR1} + v_{SGBR2} = v_{SGA} + v_{SGR}$

If $v_{SGBR1} \equiv V_{SGA}$ $\rightarrow$ Body effects match $\therefore$ Match $M_{BR1}{:}M_A$ & $M_{BR2}{:}M_R$

$\qquad\qquad \rightarrow$ $v_{SDBR1(SAT)} \equiv V_{SDA(SAT)}$ $\therefore$ $v_{SDBR2(SAT)} = v_{SDR(SAT)}$

5.5. Class-AB Op-Amp Examples: A. Cascode-Folded Cascode Followers

Bias: f(Mirrors & negative feedback)	$V_{OS(S)}$: f(Gain stage)
v_{IC} Range: f(Folded-cascode gain stage)	$V_{OS}{}^{*}$: f(Gain stage)
$v_O{:}i_O$ Range: f(Class-AB followers)	Gain:

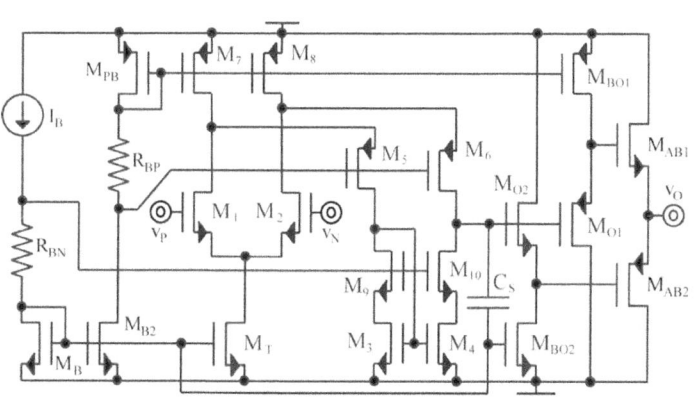

$R_{O1} = R_{D6} \parallel R_{D10}$

$A_{V0} \approx A_{V10} = g_{m12}R_{O1}$

$p_{O1} \approx 1/2\pi R_{O1}C_S$

$f_{0dB} \approx g_{m12}/2\pi C_S$

$\qquad \leq p_O \approx z_{SX} \approx g_{mB}/2\pi C_S$

$\qquad \ll p_M, z_{MX}, p_F, p_{GAB}$

PSR: Balanced N-mirror M_{34} $\rightarrow$ Followers M_{AB12} reproduce v_{ss}

B. Mirror-Folded Transconductors

Bias: f(Mirrors & negative feedback)

v_{IC} Range: f(Mirror-folded input stage)

v_O:i_O Range: f(Class-AB transconductors)

$V_{OS(S)}$:   f(Input stage)

V_{OS}^*:   f(Input stage)

Gain:

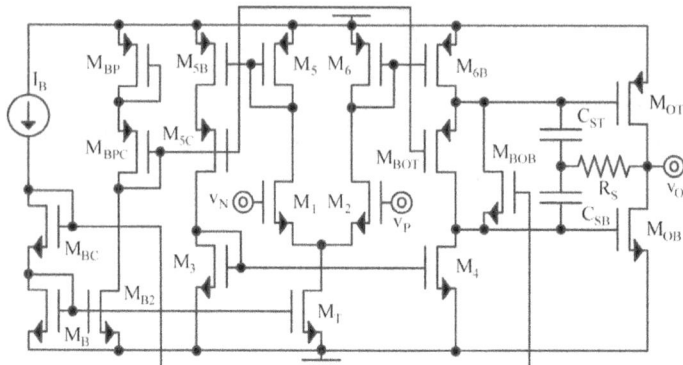

$$A_{V10} \approx -g_{m12}(S_{6B}/S_6)r_{ds46B}$$

$$A_{V20} \approx -(g_{mOT} + g_{mOB})R_O$$

$$p_{GO} \approx 1/2\pi r_{ds46B}(-A_{V20}2C_S)$$

$$f_{0dB} \approx g_{m12}(S_{6B}/S_6)/2\pi(2C_S)$$

$$\leq p_O \approx z_{SX} < p_F, p_M, z_{MX}$$

PSR: N-mirror M_{34} reproduces v_{ss} & cancels v_{dd} → M_{OT} amplifies v_{dd} & v_{ss}

5.6. Current-Mode Op Amp

Motivation for processing currents:

Low-R nodes   ∴   High-f_O poles

Low-V swings   ∴   Low slew-rate delays

                 Low charge/discharge capacitor energy

                 More headroom for supplies   →   High dynamic range

Desirable Characteristics:

Current Amplifier

High differential current gain A_I

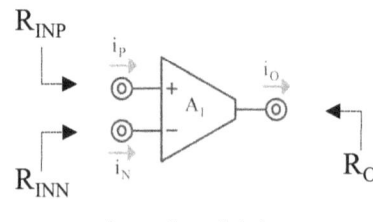

Shunt i_{IN} →   Low R_{IN}'s

Block/impede i_O →   High R_O

Source impedes i_{IN} →   High R_S

Load shunts i_O →   Low R_{LD}

$$i_O = (i_P - i_N)A_I$$

A. Concept

Idea: Accurate translation $\rightarrow$ – FB

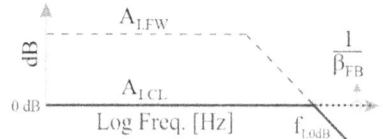

Fast $\rightarrow$ $p_{BW} = \text{Max } f_{I.0dB}$ $\rightarrow$ Min $A_{I.CL}$

$\therefore$ Unity gain $|\beta_{FB}| \equiv 1$ $\rightarrow$ $|i_{FB}| \equiv |i_O|$

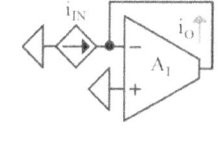

Circuit: v_{IN} to i_{IN} with R_{IN} $\rightarrow$ $\dfrac{i_{IN}}{v_{IN}} = \dfrac{1}{R_{IN}}$

i_{IN} to i_O with $A_{I.CL} \approx -1$

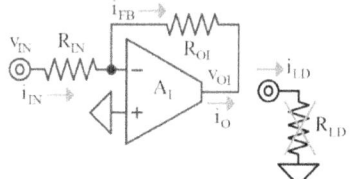

i_O to v_{OI} with R_{OI} $\rightarrow$ $\dfrac{v_{OI}}{i_O} = R_{OI}$

i_{LD} alters i_{FB} $\therefore$ No R_{LD}

$\rightarrow$ v_{OI} to v_O with high-R_{IN} buffer $A_V \approx 1$

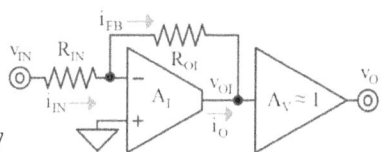

$\therefore$ $A_{I/O} = \left(\dfrac{1}{R_{IN}}\right) A_{I.CL} R_{OI} A_V$ $\rightarrow$ A_V reduces accuracy

B. Analysis

Design: R_{IN}, R_{OI} steal current from A_I's R_{IA} $\therefore$ $R_{IA} \ll R_{IN}$, R_{OI} $\therefore$ $p_{IA} \gg p_{OA}$

A_I's R_{OA} steals current from R_{OI} $\therefore$ $R_{OA} \gg R_{OI} + (R_{IN} \parallel R_{IA})$

Equations: R_{OI} $\rightarrow$ $R_I = R_{OI}$ $\qquad G_{FB} = 1/R_{OI}$ $\qquad G_{FW} = 1/R_{OI}$ $\qquad R_O = R_{OI}$

$$A_{LFW} \equiv \frac{i_O}{i_{IN} - i_{FB}} = \left(R_{IN} \parallel R_{IA} \parallel R_{OI}\right)\left[\left(\frac{-A_I}{R_{IA}}\right) + G_{FW}\right]\left(R_{OA} \parallel R_{OI}\right)G_{FB} \approx -A_I$$

$$\beta_{FB} \equiv \frac{i_{FB}}{i_O} = -1 \qquad A_{I.CL} = A_{LFW} \parallel \frac{1}{\beta_{FB}} \approx -1 \qquad f_{I.0dB} \approx A_{LFW}\beta_{FB}p_{OA} \approx A_I p_{OA}$$

$$A_{I.O} \equiv \frac{v_{IN}}{v_O} = \left(\frac{i_{IN}}{v_{IN}}\right)\left(\frac{i_O}{i_{IN}}\right)\left(\frac{v_{OI}}{i_O}\right)\left(\frac{v_O}{v_{OI}}\right) = \left(\frac{1}{R_{IN}}\right)A_{I.CL}R_{OI}A_V \approx -\frac{R_{OI}}{R_{IN}}$$

$$p_{IA} = \frac{1}{2\pi[R_{IN} \parallel R_{IA} \parallel (R_{OI} + R_{OA})]C_{IA}} \approx \frac{1}{2\pi R_{IA}C_{IA}}$$

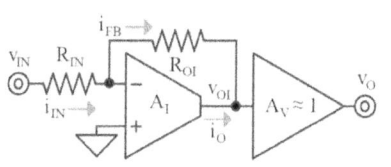

$$p_{OA} \approx \frac{1}{2\pi\{R_{OA} \parallel [R_{OI} + (R_{IN} \parallel R_{IA})]\}C_{OA}} \approx \frac{1}{2\pi R_{OI}C_{OA}}$$

85

C. Frequency Response

$$A_{I/O} \approx -\frac{R_{OI}}{R_{IN}} \text{ up to } p_{BW} = f_{I.0dB}$$

Higher $A_{I/O}$ with lower R_{IN}

→ No sacrifice in $p_{BW} \neq f(R_{IN})$

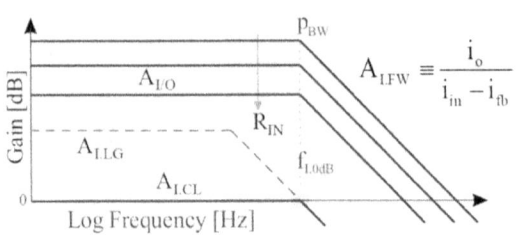

$$A_{I.FW} \equiv \frac{i_o}{i_{in} - i_{fb}}$$

D. Example

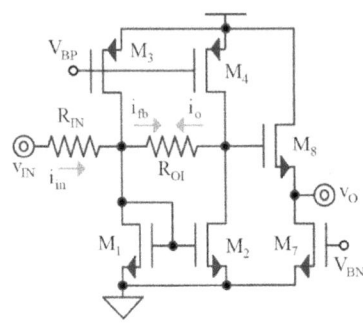

$$R_{IA} = \frac{1}{g_{m1}} \| r_{ds1} \| r_{ds3} \qquad\qquad R_{OA} = r_{ds2} \| r_{ds4}$$

$$C_{IA} \approx C_{GS1} + C_{GS2} \qquad\qquad C_{OA} \approx C_{GD2} + C_{GD4} + C_{GD8}$$

$$A_{I.FW} = (R_{IN} \| R_{IA} \| R_{OI})(-g_{m2})\left(\frac{R_{OA} \| R_{OI}}{R_{OI}}\right) \approx -\frac{g_{m2}}{g_{m1}}$$

$$A_V = \frac{g_{m8}(r_{ds7} \| R_{LD})}{1 + (g_{m8} + 1/r_{ds8})(r_{ds7} \| R_{LD})} \text{ with } p_O, z_{GS8}$$

Reduce R_{IA}: Raise W_{12} or I_{34} Raise R_{OA}: Lengthen L_{1-4} or degenerate/cascode M_{24}

Chapter 6. Linear Voltage Regulators

6.1. Functionality

6.2. Frequency Response

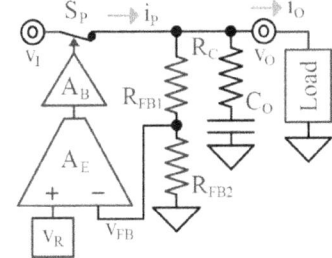

6.3. Power-Supply Rejection

6.4. IC Design

6.1. Functionality: A. Parameters

Purpose: Supply P_O

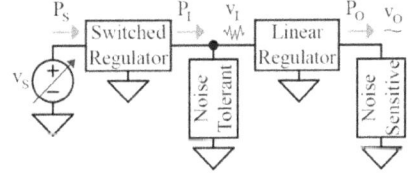

Suppress v_I noise v_i

"Regulate" v_O near target $v_O{}'$

Accuracy:	Power:	Operation:
v_I–i_O–T_J Sensitivity	$i_{O(MAX)}$	Range limits:
V_{OS}	P_{GND} Loss	v_O, v_I, i_O, T_J, C_O
Response Time t_R	P_{DROP} Loss	Related Terms:
Implied & Other Specifications:		Line (v_I) Regulation
A_{LG}	i_Q	Load (i_O) Regulation
R_O	Drop $\equiv v_I - v_O$	Load-Dump Response
f_{BW}	Cost...	Temperature (T_J) Drift

B. Output Regulation

How: – FB loop ensures $v_O \approx \beta_{FB}$ translation of accurate v_R $\therefore$ $v_O \neq f(v_I, i_O)$

Composition:

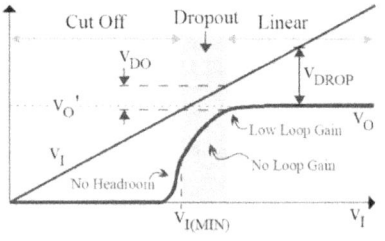

Pass transistor S_P $\rightarrow$ Draw & supply i_P

FB translation β_{FB} $\rightarrow$ Scale v_O to v_{FB}

Reference v_R $\rightarrow$ Insensitive to v_I & T_J

Error Amplifier A_E $\rightarrow$ Sense & amplify error $v_E = v_R - v_{FB}$

Buffer A_B $\rightarrow$ Unload C_{PI} from R_{EO}

Drive C_{PI} with low R_{BO}

Output Filter C_O $\rightarrow$ Hold v_O & supply sudden Δi_{LD}

$R_C \equiv$ Parasitic equivalent series resistance (ESR)

Operating Regions

Linear: $A_{LG} \gg 1$ $\therefore$ $v_O = \dfrac{v_{FB}}{\beta_{FB}} \approx \dfrac{v_R}{\beta_{FB}}$ $\left[\begin{array}{l} \text{Low } v_I \text{ Sensitivity} \quad \rightarrow \quad \text{Low } \Delta v_{LN} \\[1em] \text{Low } i_O \text{ Sensitivity} \quad \rightarrow \quad \text{Low } \Delta v_{LD} \end{array} \right.$

Dropout: $A_{LG} \leq 1$ $\therefore$ v_O shifts $\rightarrow$ Drops out of regulation

When feedback transistor(s) enter(s) triode

$v_{DO} \equiv v_I - v_O$ when $A_{LG} = 1$

$P_{PO} = i_{PO}(v_I - v_O) \geq i_{PO}v_{DO}$

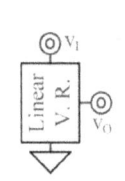

$\rightarrow$ Dropout is least lossy, linear is most accurate $\therefore$ Best just above dropout

Cut off: $v_I <$ Headroom $v_{I(MIN)}$ $\therefore$ i_{LD} pulls v_O towards ground

C. Offset

Cause: V_{EO} needed to supply I_P → $f(i_{LD})$ ∴ Loading effect

$$V_{OS} \equiv V_R - V_{FB} = \frac{V_{EO} - V_{EO0}}{A_E}$$ $V_{EO0} \equiv A_E\text{'s Zero-Offset Level}$

Center: If possible, set V_{EO0} near V_{EO}'s center → $V_{EO0} \approx \dfrac{V_{EO(LD1)} + V_{EO2(LD2)}}{2}$

Compensate: V_{OS} reduces v_{FB} ∴ Adjust & center $\beta_{FB} \equiv \dfrac{\overline{V_{FB}}}{V_O} = \dfrac{v_R - \overline{V_{VOS}}}{V_O}$

Load Regulation:

FB model → 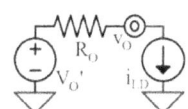 $R_O = \dfrac{1}{G_{LG}}$ ∴ $\Delta v_{LD} = \Delta i_{LD} R_O$

Target
↑
Centered/Compensated Output: $V_O \approx V_O' \pm 0.5\Delta v_{LD}$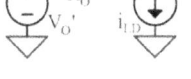

D. Load-Dump Response

RC Step Response: $2.3\tau_{RC}$ to 90% of target   → $t_R \approx \dfrac{2.3}{2\pi f_{0dB}} = \dfrac{37\%}{f_{0dB}}$

Load Dump: $\Delta i_{LD} \equiv$ Wide & sudden i_{LD} variation = f(Application)

v_O Response: i_P' needs t_R to supply new i_{LD} → C_O & C_{LD} supply $\pm\Delta i_{LD}$ across t_R

R_C drops $\Delta v_R = i_C R_C$

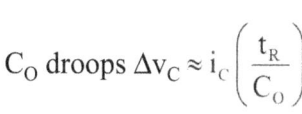 $i_C \approx \Delta i_{LD} \left(\dfrac{C_O}{C_O + C_{LD}} \right)$

C_O droops $\Delta v_C \approx i_c \left(\dfrac{t_R}{C_O} \right)$

$i_P' - i_{LD}$ re/discharges C_O & C_{LD} across t_{SR}

Usually, $t_R^+ = f(i_{O(MIN)}) > t_R^- = f(i_{O(MAX)})$

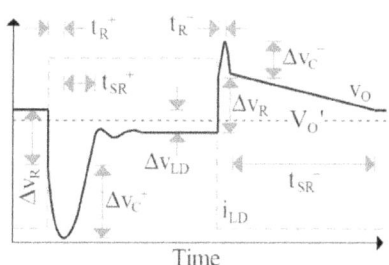

Accuracy: $\Delta v_{O(MAX)} = \Delta v_R + \Delta v_{C(MAX)} - 0.5\Delta v_{LD}$

Example: Determine Δv_{LD}, Δv_R, Δv_C, & $\Delta v_{O(MAX)}$ when Δi_{LD} = 100 mA in 1 ns,

v_O' = 2 V, C_O = 1 µF, R_C = 250 mΩ, C_{LD} = 100 nF, f_{0dB} = 500 kHz, R_O = 300 mΩ.

Solution:

$$\Delta v_{LD} = \Delta i_{LD} R_O = 30 \text{ mV} \quad \rightarrow \quad 1.5\% v_O' \qquad\qquad t_R \approx 2.3 \tau_{BW} = \frac{2.3}{2\pi f_{0dB}} = 730 \text{ ns}$$

$$\Delta v_R \approx \left(\frac{\Delta i_{LD} C_O}{C_O + C_{LD}} \right) R_C = 23 \text{ mV} \quad \rightarrow \quad 1.1\% v_O'$$

$$\Delta v_C \approx \Delta i_{LD} \left(\frac{t_R}{C_O + C_{LD}} \right) \approx 66 \text{ mV} \quad \rightarrow \quad 3.3\% v_O'$$

$$\Delta v_{O(MAX)} \approx \Delta v_R + \Delta v_C - 0.5 \Delta v_{LD} = 74 \text{ mV} = 3.7\% v_O' \quad \rightarrow \quad \Delta v_{LD} \text{ reduces } \Delta v_{O(MAX)}$$

E. Power Consumption

Ohmic Drop $P_{DROP} \equiv i_O v_{DROP} = i_O (v_I - v_O) = i_O$ part of S_P power (without i_{FB})

Ground Loss $P_{GND} = i_{GND} v_I = (i_{FB} + i_Q) v_I$

Power-Conversion Efficiency:

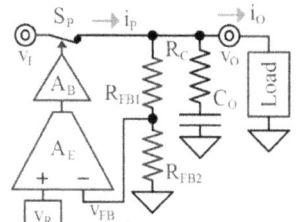

$\eta_C \equiv$ Fraction of P_I that P_O delivers

$$\eta_C \equiv \frac{P_O}{P_I} = \frac{i_O v_O}{i_I v_I} = \frac{i_O v_O}{(i_O + i_{GND}) v_I} = \eta_I \left(\frac{v_O}{v_I} \right) < \frac{v_O}{v_I}$$

$$\underset{\text{Current Efficiency}}{\downarrow}$$

$$\eta_C = \eta_I \left(\frac{v_I - v_{DROP}}{v_I} \right) < \eta_I \left(\frac{v_I - v_{DO}}{v_I} \right) = \eta_I \left(1 - \frac{v_{DO}}{v_I} \right)$$

6.2. Frequency Response: A. Feedback Model

v_i: Does not respond to loop signals   →   No small signals   → $v_i = 0$

A_E: Norton equivalent   → $R_{EI} : C_{EI}$ G_E $R_{EO} : C_{EO}$

A_B: Norton equivalent   → $R_{BI} : C_{BI}$ G_B $R_{BO} : C_{BO}$

S_P: Norton equivalent   → $R_{PI} : C_{PI}$ G_P $R_{PO} : C_{PO}$

Model: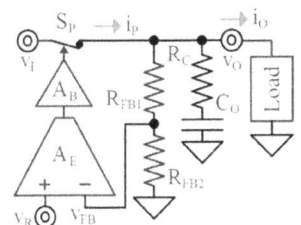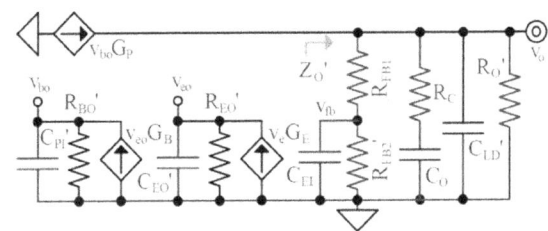

Combinations: $R_O' \equiv R_{LD} \| R_{PO}$ $C_{LD}' \equiv C_{LD} \| C_{PO}$ $R_{FB2}' \equiv R_{FB2} \| R_{EI}$

$R_{EO}' \equiv R_{EO} \| R_{BI}$ $C_{EO}' \equiv C_{EO} \| C_{BI}$ $R_{BO}' \equiv R_{BO} \| R_{PI}$ $C_{PI}' \equiv C_{BO} \| C_{PI}$

B. Relative Values

C_O supplies Δi_{LD} across t_R so $\Delta v_O =$ Low   ∴   High C_O

S_P supplies P_O ∴   Large S_P →   Wide W, Short L   →   Moderate C_{IP}, R_{PO}

R_{FB1} & R_{FB2}' pull low $i_{FB} = \dfrac{v_O}{R_{FB1} + R_{FB2}'} = \dfrac{v_{FB}}{R_{FB2}'} \approx \dfrac{v_R}{R_{FB2}}$ ∴   High $R_{FB1} + R_{FB2}'$

A_E amplifies v_E →   Low C_{EI} High $A_E = G_E R_{EO}'$ ∴   Very high R_{EO}'

A_B buffers v_{eo} →   Low C_{BI} Low R_{BO} ∴   Low A_B

Large load adds moderate C_{LD}

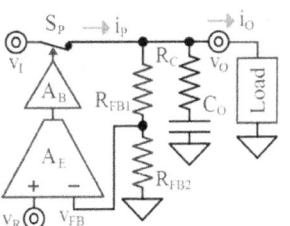

Result: $R_{EO} \geq R_{FB1} + R_{FB2}' >> R_{PO} \| R_{LD} >> R_C$

       $C_O >> C_{LD} >> C_{PI} >> C_{EI}, C_{BI}$

C. Frequency Response

Loop Gain: $A_{LG} \equiv \dfrac{s_{FB}}{s_E} = A_E A_B A_P \beta_{FB} = G_E Z_{EO} G_B Z_{BO} G_P Z_O' \left(\dfrac{Z_{FB2}}{R_{FB1} + Z_{FB2}} \right)$

Low-f_O Gain: $A_{LG0} = G_E R_{EO}' G_B R_{BO}' G_P [(R_{FB1} + R_{FB2}') \| R_O'] \left(\dfrac{R_{FB2}}{R_{FB1} + R_{FB2}'} \right)$

$$\approx G_E R_{EO} G_B R_{BO} G_P \left(R_{PO} \| R_{LD} \right) \left(\dfrac{R_{FB2}}{R_{FB1} + R_{FB2}} \right)$$

f_O Response: Z_{EO}: C_{EO}' shunts R_{EO}' past $p_E = \dfrac{1}{2\pi R_{EO}' C_{EO}'}$

Z_{BO}: C_{PI}' shunts R_{BO}' past $p_B = \dfrac{1}{2\pi R_{BO}' C_{PI}'}$

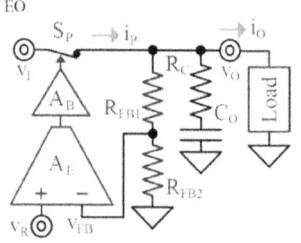

G_P: Cross-S_P C_{PX} overcomes G_{P0} past $z_P = \dfrac{G_{P0}}{2\pi C_{PX}}$

Z_O': Low f_O: $Z_{CO} + R_C \approx Z_{CO}$ $C_O \gg C_{EI}$

$\therefore$ $C_O + C_{LD}'$ shunts $(R_{FB1} + R_{FB2}') \| R_O'$ past:

$$p_O = p_C \approx \dfrac{1}{2\pi \left[\left(R_{FB1} + R_{FB2}' \right) \| R_{PO} \| R_{LD} \right] \left(C_O + C_{LD} + C_{PO} \right)}$$

Higher f_O: $C_{LD}' \gg C_{EI}$ R_{FB1}, R_{FB2}', and R_O' fade

→ R_C current-limits C_O past $z_{CX} = \dfrac{1}{2\pi R_C C_O} > p_O$

C_{LD}' shunts R_C past $p_{LD} \approx \dfrac{1}{2\pi R_C \left(C_{LD} + C_{PO} \right)} > z_{CX}$

β_{FB}: $R_{FB1} \gg Z_{LD}'$ → C_{EI} shunts $R_{FB1} \| R_{FB2}'$ past $p_{FB} \approx \dfrac{1}{2\pi \left(R_{FB1} \| R_{FB2}' \right) C_{EI}}$

D. Stabilization

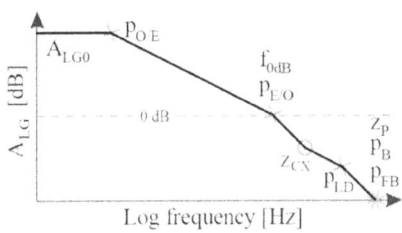

Stabilized Response:

A_{LG} drops after p_O or p_E

$$\therefore \quad p_{O/E} \ll f_{0dB} \leq p_{E/O} \; z_{CX} - p_{LD} < p_B \; z_P \; p_{FB}$$

Challenges:

$R_{LD} \approx \dfrac{v_O}{i_O}$ for resistive loads, $R_{LD} \to \infty$ for active loads $\rightarrow$ Wide R_{LD} range

$i_O = \mu A's$ to $mA's$ $\quad \rightarrow \quad p_O = f(R_O') = f(R_{PO}) = f(i_O)$ $\rightarrow$ Widely variable p_O

$C_O = nF's$ to $\mu F's$ $\quad \rightarrow \quad$ Low-to-moderate p_O

$A_E \geq 20$ dB $\quad\quad \rightarrow \quad$ Moderate-to-very high R_{EO} $\therefore$ Low-to-moderate p_E

$C_{PI} \approx pF's$ $\quad\quad\quad \rightarrow \quad$ R_{BO} should be low $\quad\quad \therefore$ Requires i_Q

$R_C \leq 1 \; \Omega$ $\quad\quad\quad \rightarrow \quad$ High-to-very high $z_{CX} - p_{LD}$ $\therefore$ z_{CX} does not help

Example: Determine C_O and R_{BO}' so $f_{0dB} \approx p_E$ and PM $\approx 30°$ when $G_E = 100 \; \mu S$,

$R_{EO}' = 1 \; M\Omega$, $C_{EO}' = 100 \; fF$, $A_B = 1 \; V/V$, $G_P = 50 \; mS$, $C_{PI}' = 10 \; pF$,

$R_O' = 100 \; \Omega$, $R_{FB1} = R_{FB2}' = 250 \; k\Omega$, $R_C = 200 \; m\Omega$, $C_{LD}' = 1 \; nF$, $C_{EI} = 50 \; fF$.

Solution:

$$p_E = \frac{1}{2\pi R_{EO}' C_{EO}'} = 1.6 \; MHz$$

$$f_{0dB} \approx GBW \approx G_E R_{EO}' A_B G_P R_O' \left(\frac{R_{FB2}'}{R_{FB1} + R_{FB2}'} \right) p_O \equiv p_E \quad \therefore \quad p_O = 6.4 \; kHz$$

$$p_O \approx \frac{1}{2\pi \left[(R_{FB1} + R_{FB2}') \| R_O' \right](C_O + C_{LD}')} \equiv 6.4 \; kHz \quad \therefore \quad C_O = 250 \; nF$$

$$z_{CX} = \frac{1}{2\pi R_C C_O} = 3.2 \text{ MHz}$$

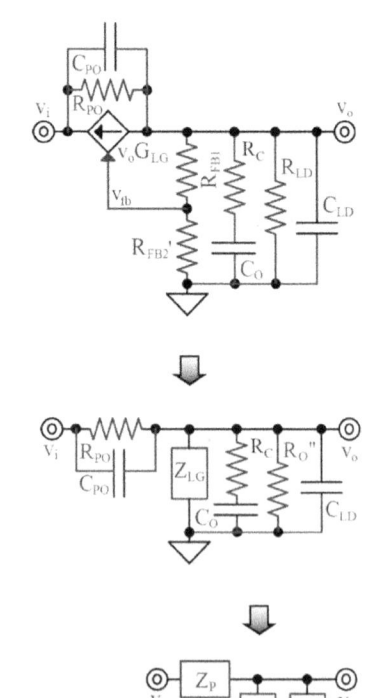

$$p_{FB} \approx \frac{1}{2\pi \left(R_{FB1} \parallel R_{FB2}' \right) C_{EI}} = 25 \text{ MHz}$$

$$p_{LD} \approx \frac{1}{2\pi R_C C_{LD}'} = 320 \text{ MHz}$$

$$\angle A_{LG}\Big|_{f_{0dB}} = \overset{-90°}{-\tan^{-1}\left(\frac{f_{0dB}}{p_O}\right)} \overset{-45°}{-\tan^{-1}\left(\frac{f_{0dB}}{p_E}\right)} -\tan^{-1}\left(\frac{f_{0dB}}{p_B}\right) \overset{+27°}{+\tan^{-1}\left(\frac{f_{0dB}}{z_{CX}}\right)} \overset{-4°}{-\tan^{-1}\left(\frac{f_{0dB}}{p_{FB}}\right)}$$

$$\overset{-0°}{-\tan^{-1}\left(\frac{f_{0dB}}{p_{LD}}\right)} = -112° -\tan^{-1}\left(\frac{f_{0dB}}{p_B}\right) \equiv -180° + \overset{30°}{PM} \equiv -150°$$

$$\rightarrow \quad p_B = \frac{1}{2\pi R_{BO}' C_{PI}'} \geq 2.1 \text{ MHz} \quad \therefore \quad R_{BO}' \leq 7.6 \text{ k}\Omega$$

6.3. Power-Supply Rejection: A. Model

Assumption: $A_E : A_B : S_{PO}$ do not inject v_i noise

Voltage Divider: $\quad R_O'' \equiv (R_{FB1} + R_{FB2}') \parallel R_{LD}$

$\quad\quad\quad\quad\quad\quad\quad G_{LG}$'s shunt-FB Z_{LG} shunts v_O

$$Z_{LG} \equiv \frac{v_o}{i_{GP}} = \frac{1}{G_{LG}} = \frac{1}{\beta_{FB} A_E A_B G_P}$$

$$A_I \equiv \frac{v_o}{v_i} = \frac{Z_G}{Z_P + Z_G} = \frac{Z_P \parallel Z_G}{Z_P} = \frac{Z_P \parallel Z_{LG} \parallel Z_O''}{Z_P}$$

PSR $\equiv$ Inability to amplify $v_i \equiv \dfrac{1}{A_I} = \dfrac{v_i}{v_o}$

Line Regulation: $A_{LN} = \dfrac{\Delta v_O}{\Delta v_I}\bigg|_{DC} \equiv \dfrac{\Delta v_{LN}}{\Delta V_I} \approx \dfrac{v_o}{v_i}\bigg|_{DC} = A_{I0} = \dfrac{1}{PSR_0}$

B. Frequency Response

$$A_{I0} = \frac{R_{PO} \parallel R_{LG} \parallel R_O"}{R_{PO}} \approx \frac{R_{LG}}{R_{PO} + R_{LG}} \approx \frac{1}{G_{LG0}R_{PO}} \approx \frac{1}{A_{LG0}}$$

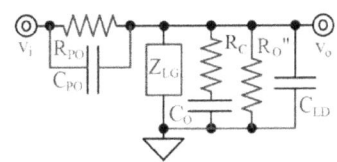

Z_{LG}: p_E raises $Z_{LG} = \dfrac{1}{G_{LG}}$ → A_I rises past z_E'

Until Z_{LG} overcomes $Z_{PO} \parallel Z_O"$ → z_E reverses past $p_{EX}' = f_{0dB}$

If $p_E \ll f_{0dB} < p_O < f_{CP}$: $\dfrac{1}{G_{LG}} \geq R_{PO} \parallel R_O"$ → $A_{LG} \leq 1$ ∴   Past f_{0dB}

If $p_O \ll p_E < f_{0dB} < f_{CP}$: $\dfrac{1}{G_{LG}} \geq Z_{CO}"$ →   Past $\underbrace{G_{LG}(R_{PO} \parallel R_O")p_O = f_{0dB}}_{GBW}$

C_O: $C_O"$ shunts $R_{PO} \parallel Z_{LG} \parallel R_O"$ → A_I falls past p_O or p_O'

If $p_E \ll f_{0dB} < p_O < f_{CP}$: Z_{LG} fades past f_{0dB} ∴ $Z_{CO}" \leq R_{PO} \parallel R_O"$ past p_O

If $p_O \ll f_{0dB} < p_E, f_{CP}$: $Z_{CO}" \leq \dfrac{1}{G_{LG0}}$ → Past $\underbrace{G_{LG0}(R_{PO} \parallel R_O")p_O = f_{0dB} = p_O'}_{GBW}$

R_C current-limits (shorts) C_O past p_O or p_O' → p_O or p_O' reverses past z_{CX}

C_{PO}: C_{PO} bypasses R_{PO} →   Couples v_i past z_{PO}

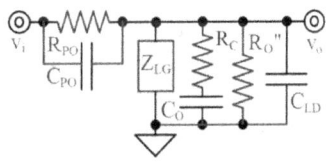

C_{LD}: C_{LD} shunts $Z_{PO} \parallel R_C \parallel R_O"$ past z_{CX} →   Shunts v_i past p_{LD}

When $S_P \equiv$ P-Channel MOSFET:

C_{GD} diode-connects M_{PO} $\rightarrow$ $Z_{GP} = \dfrac{1}{G_{PO}} = \dfrac{Z_{GD} + \left(Z_{GS} \| R_{BO}\right)}{\left(Z_{GS} \| R_{BO}\right)g_{mPO}} \rightarrow \infty$ At low f_O

Z_{GP} falls as C_{GD} shunts with f_O:

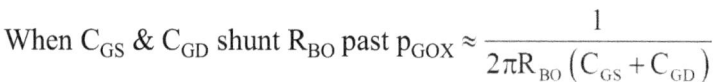

Z_{GP} bypasses r_{dsPO} when $Z_{GP0} = \dfrac{Z_{GD}}{R_{BO}g_{mPO}} \leq r_{dsPO}$

past $z_{GO} = \dfrac{1}{2\pi R_{BO}g_{mPO}r_{dsPO}C_{GD}} = \dfrac{1}{2\pi R_{BO}g_{mPO}R_{PO}C_{GD}} \approx p_B$

$z_B{}'$ reverses when Z_{GP} reaches $Z_{GP(HF)} = \dfrac{C_{GD} + C_{GS}}{C_{GD}g_{mPO}}$

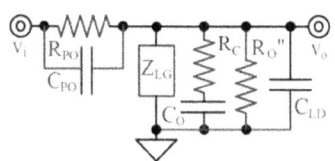

When C_{GS} & C_{GD} shunt R_{BO} past $p_{GOX} \approx \dfrac{1}{2\pi R_{BO}\left(C_{GS} + C_{GD}\right)}$

C. Internally Stabilized Example

$A_{LN} = A_{I0} = \dfrac{R_{PO} \| R_{LG} \| R_O{}''}{R_{PO}} \approx \dfrac{R_{LG}}{R_{PO} + R_{LG}}$

Z_{LG} rises past p_E

$Z_{LG} \geq R_{PO} \| R_O{}''$ past f_{0dB}

$Z_{CO}{}'' \leq R_{PO} \| R_O{}''$ past p_O

$Z_{CP} \leq R_{PO}$ past z_{PO}

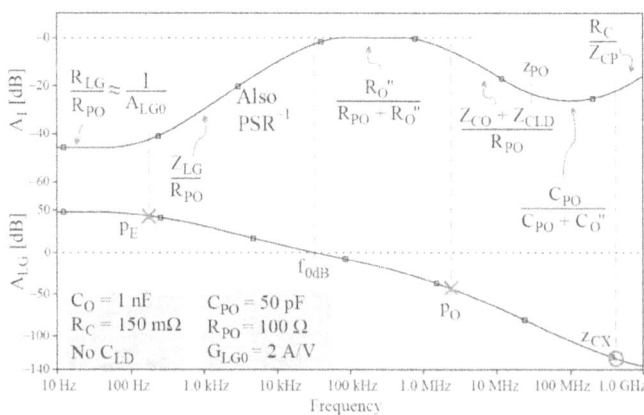

R_C shorts C_O past z_{CX}

A_I past f_{0dB}, below p_O, $z_{PO} \approx \dfrac{R_O{}''}{R_{PO} + R_O{}''}$

A_I past p_O, $z_{PO} \approx \dfrac{C_{PO}}{C_{PO} + C_O{}''}$

D. Output Stabilized Example

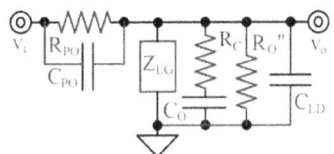

$$A_{LN} = A_{I0} = \frac{R_{PO} \| R_{LG} \| R_O"}{R_{PO}} \approx \frac{R_{LG}}{R_{PO} + R_{LG}}$$

$Z_{CO}"$ shunts $R_{PO} \| R_O"$ past p_O Z_{LG} rises past p_E

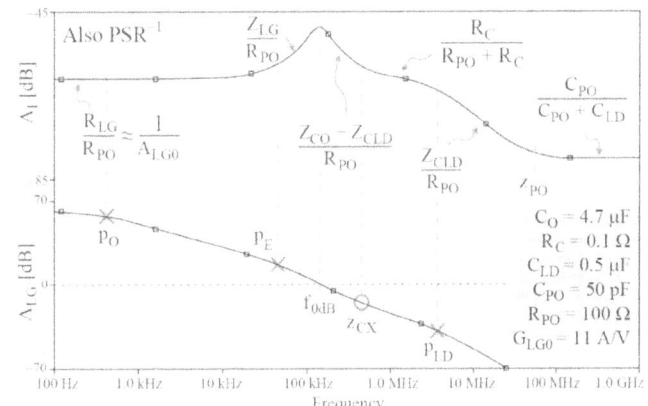

$Z_{LG} \geq Z_{CO}"$ past f_{0dB}

$Z_{CO}"$ falls past $p_O' = f_{0dB}$

R_C shorts C_O past z_{CX}

Z_{CLD} shunts $R_{PO} \| R_C$ past p_{LD}

A_I past z_{CX}, below p_{LD}, $z_{PO} \approx \dfrac{R_C}{R_{PO} + R_C}$ A_I past p_{LD}, $z_{PO} \approx \dfrac{C_{PO}}{C_{PO} + C_{LD}}$

D. Conclusions

$$PSR = \frac{1}{A_I} \quad \rightarrow \quad \text{Derive } A_I$$

$$A_{LN} = A_{I0} = \frac{1}{PSR_0} \approx \frac{1}{A_{LG0}}$$

R_{PO}, C_{PO} couple v_i → Increase R_{PO} Decrease C_{PO}

Z_{LG}, C_O, C_{LD} shunt v_o → Decrease Z_{LG} Increase C_O, C_{LD}

$$R_{LG} = \frac{1}{G_{LG0}} \quad \rightarrow \quad \text{Increase } G_{LG0}$$

$p_{E/B}$ reduce G_{LG} (raise Z_{LG}) → Keep over f_{0dB} ∴ Stabilize with p_O when possible

Z_{LG} shunts v_o up to f_{0dB} → PSR $\neq$ f(FB) past f_{0dB}

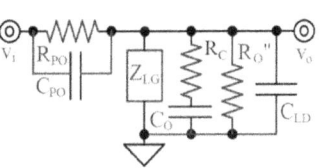

R_C current-limits C_O → Choose C_O with lower R_C

Design Assumption: G_{LG} does not inject v_i

E. High-PSR Strategies

Filter v_I: In-line RC filter → Lossy ∴ Small R_F → Suppress high-f_O v_i

In-line regulator → Suppress up-to-f_{0dB} v_i, lossy ∴ Low v_{DO}

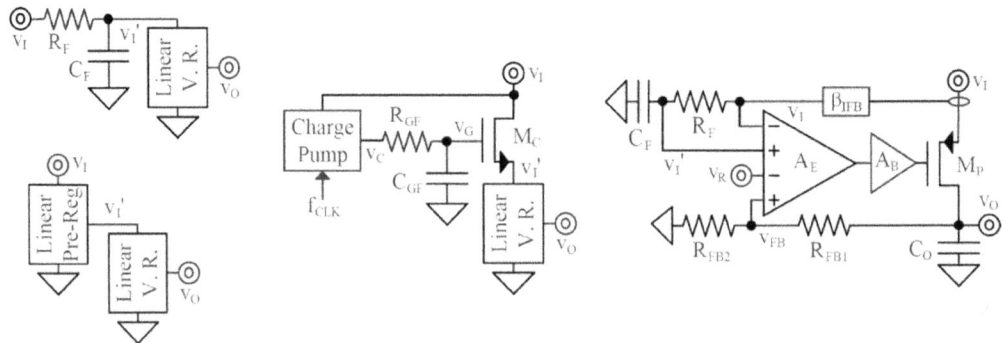

Raise R_I: Cascode → Charge-pump, filter v_G → Suppress all-f_O v_i, add v_{DROP}

Series FB → i-mode voltage loop → Suppress about-f_{0dB} v_i

6.4. IC Design: Design Process

Design: From specified targets

Typical Design Process

C_O: From Δv_O, Δi_{LD}, & PSR

β_{FB}: From V_O', v_R, & i_{GND}

S_O: Type from η_C & t_R

Size from v_{DO} & i_O

A_B: From S_O requirements

A_E: From A_B & S_O requirements

A. Power Pass Transistor: N Type

Dropout:

$$v_{NDO(BJT)} > v_{BE} \approx V_t \ln\left(\frac{i_L}{I_S}\right)$$

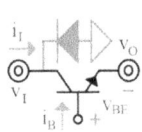

$$v_{NDO(MOS)} > v_{GS} \approx V_{TN0} + \gamma_N \left(\sqrt{2\psi_B + v_O} - \sqrt{2\psi_B}\right) + \sqrt{\frac{2i_{IN}}{K_N'(W/L)}} \quad \rightarrow \quad \text{High W/L}$$

$$\overset{\text{High}}{\therefore \;\; v_{NDO} \geq 500 \text{ mV}} \quad \rightarrow \quad \text{High-dropout (HDO) regulator} \quad \rightarrow \quad \text{Lossy}$$

NPN: v_I-fed $i_B = \dfrac{i_L}{\beta_N}$ flows to v_O $\rightarrow$ $i_B \neq$ Not lost

R's: $R_{PI(NPN)} \approx r_\pi + R_{LD} + g_m r_\pi R_{LD}$ $\qquad R_{PO} \approx \dfrac{1}{g_m'}$ $\qquad$ Short L $\rightarrow$ Noticeable λ

Follower Response: Δv_O $\rightarrow$ $\Delta v_{BE/GS}$ $\rightarrow$ Δi_I $\quad \therefore$ Short t_R $\rightarrow$ Fast (–FB)

P Type

Dropout:

$$v_{PDO(BJT)} = v_{EC(MIN)}$$

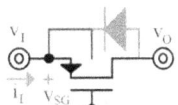

$$v_{PDO(MOS)} = v_{SD(TRI)} \geq i_L R_{CH} = \frac{i_L}{K_P'(W/L)\left(v_{SG} - |V_{TP0}| - 0.5 v_{DO}\right)} \quad \rightarrow \quad \text{High W/L}$$

$$\therefore \;\; v_{PDO} \approx 200\text{--}300 \text{ mV} \quad \rightarrow \quad \text{Low-dropout (LDO) regulator} \quad \rightarrow \quad \text{Higher } \eta_C$$

PNP: i_{IN} loses $i_B \approx \dfrac{i_L}{\beta_P}$ to ground $\rightarrow$ PNP = Lossier than PMOS

R's: $R_{PI(PNP)} = r_\pi$ $\qquad R_{PO} = R_{CH} \rightarrow r_{o/ds}$ $\qquad$ Short L $\rightarrow$ Noticeable λ

Transconductor Response: Wait for A_E to respond $\rightarrow$ Longer t_R than N type

B. Buffer

Requirement: High p_B → Low R_{BO}

Voltage Follower: $R_{BO} \approx \dfrac{1}{g_{mB}}$ $p_B \approx \dfrac{g_{mB}}{2\pi C_{PI}'}$

Substrate MOS: Body effect → $v_T > V_{T0}$

P-Type: Higher v_{BO} $i_{BO(MAX)} \le I_B$

N-Type: Lower v_{BO} $i_{BO(MIN)} \le I_B$

Native NMOS: No v_{TN} adjust implant → Low V_{TN0}

Isolated Sub. NMOS: With deep N^+ plugs & buried layer → $v_T = V_{TN0}$

Looped Follower: $i_{BO(MAX)} = i_{FB(MAX)} \ne f(I_B)$ → On demand ∴ Higher η_I

$$G_{BLG0} \approx (-g_{mB})r_{dsBN}(-g_{mFB}) = g_{mB}r_{dsBN}g_{mFB}$$

Reverse Loop:

$$G_{BLG0} = (-g_{mB})R_{EC}\left(\frac{1}{R_{GMC}}\right)r_{dsBP}(-g_{mFB}) \approx g_{mB}r_{dsBP}g_{mFB}$$

$$A_{BLG0} = \frac{v_{bo}}{v_{eo} - v_{bo}} \approx G_{BLG0}(R_{SB} \parallel R_{PI}) \qquad p_{BS} \approx \frac{1}{2\pi r_{dsBN\,P}C_{BS}}$$

∴ $R_{BO} \approx \dfrac{1}{G_{BLG0}}$ $p_B = f_{B0dB} \approx A_{BLG0}p_{BS}$

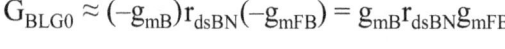
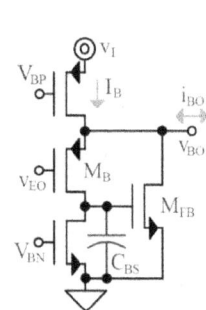
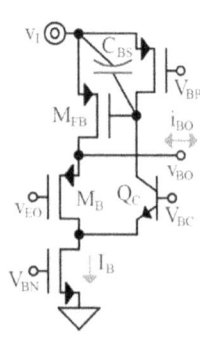

C. Design Examples: Internally Stabilized NPN

$S_O \equiv$ NPN $\rightarrow$ Fast response, Low R_{NO} $\therefore$ $p_E \ll f_{0dB}$ p_O z_S $p_B < p_{FB}$ $z_{CX}-p_{LD}$

$v_R = 1.2$ V $\rightarrow$ N-type diff. pair   High v_{BO} $\rightarrow$ P-type follower

Cancel v_i $\rightarrow$ Fold into N-type mirror   Supply i_{BO} $\rightarrow$ Reverse-loop follower

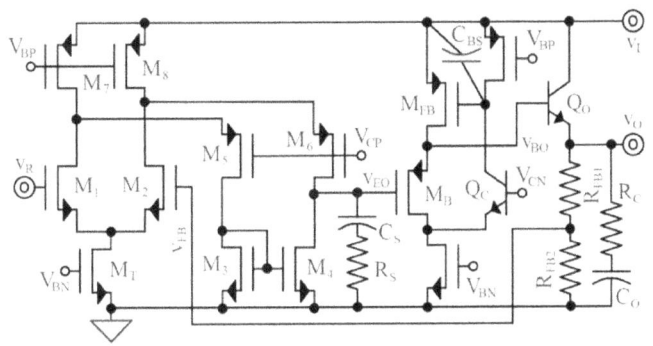

C_S shunts v_{EO} past

$$p_E = \frac{1}{2\pi (R_{EO} + R_S) C_S}$$

R_S current-limits C_S

past $z_S = z_{EX} = \dfrac{1}{2\pi R_S C_S}$

Output Stabilized PNP

$S_O \equiv$ PNP $\rightarrow$ Low dropout, Moderate R_{PO}, High PSR $\therefore$ $p_O \ll p_E$

$v_R = 1.2$ V $\rightarrow$ N-type diff. pair   Shut with high v_{BO} $\rightarrow$ P-type follower

Reproduce v_i $\rightarrow$ P-type mirror   Sink i_{BO} $\rightarrow$ Looped P-type follower

Low $v_{I(MIN)}$ $\rightarrow$ Reverse-fold into mirror C_S bypasses R_{FB1} past

$$z_S = \frac{1}{2\pi R_{FB1} C_S}$$

$p_O \ll z_S \rightarrow Z_{CO} \ll R_{FB12}$

C_S shorts past $p_{FB} = p_{SX}$

$$p_{SX} \approx \frac{1}{2\pi (R_{FB1} \| R_{FB2}) C_S}$$

Stability: $p_O \ll z_S$ f_{0dB} p_E $p_{SX} < p_B$ $z_{CX}-p_{LD}$

Internally Stabilized PMOS

$S_O \equiv$ PMOS $\rightarrow$ Low dropout, Low i_{GND}, Moderate R_{PO} $\therefore$ Split $p_E : p_O$

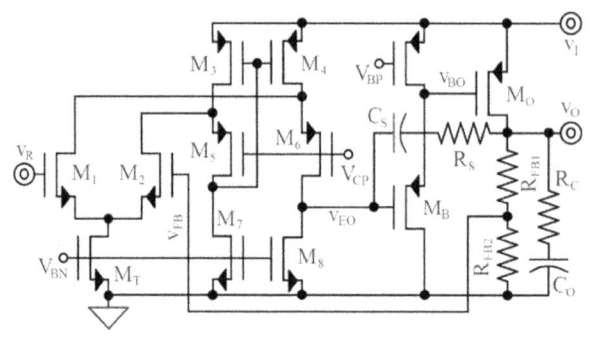

C_{SI} shunts R_{EO} past

$$p_E \approx \frac{1}{2\pi R_{OE} G_{PO}\left(R_{PO} \| R_{LD}\right) C_S}$$

R_S current-limits C_S past

$$z_{EX} = \frac{1}{2\pi\left(R_S - 1/G_{PO}\right)C_S}$$

Stability: $p_E \ll f_{0dB}$ p_O z_{EX} p_B $<$ p_{FB} z_{CX} p_{LD} $\therefore$ p_E raises A_I below f_{0dB}

Capless: On-chip C_O when load dumps are light

Buffer: M_{PB}'s I_{BP} slews M_O's C_{GO} $\rightarrow$ Reverse-loop follower to increase η_I

Example: Design S_P so $v_{DO(MIN)} = 200$ mV, $i_{GND} = 0$ when $v_O = 1.8$ V, $i_O = 50$ mA,

$v_{IN/SG} = 2$ V, $V_{TP0} = -400$ mV, $K_P' = 40$ µA/V^2, $\lambda_P = 10\%$ V^{-1}, $L \geq 180$ nm, $L_{OL} = 30$ nm.

Solution: Above dropout $\rightarrow$ $v_{SD} > v_{DO(MIN)}$ $\rightarrow$ Lower λ_P effects in g_m & $v_{SD(SAT)}$

Low v_{DO} ⎫
⎬ P-Type MOSFET
$i_{GND} = 0$ ⎭ Low R_{CH} $\therefore$ $L \equiv L_{MIN}$ $\rightarrow$ $L_{CH} = L_{MIN} - 2L_{OL} = 120$ nm

$$v_{DO(MIN)} = i_{IN}R_{CH} \approx \frac{i_O}{K_P'\left(W/L_{CH}\right)\left(v_{IN} - |V_{TP0}| - 0.5v_{DO(MIN)}\right)} \equiv 200 \text{ mV} \quad \therefore \quad W \equiv 500 \text{ µm}$$

$$v_{SD(SAT)} \approx \sqrt{\frac{2i_O}{K_P'(W/L)}} \leq 780 \text{ mV} \quad \rightarrow \quad R_{PO} \approx r_{ds} \text{ when } v_{IN} \geq v_O + v_{SD(SAT)} \approx 2.6 \text{ V}$$

$\therefore$ R_{PO} falls when $v_{IN} \leq 2.6$ V $\rightarrow$ $R_{PO} \approx R_{CH}$ when $v_{IN} \approx 2$ V

$\quad\quad\quad\quad$⎣$\rightarrow$ "Starts dropping out"

Gate-Coupled LDO

$$A_P \approx G_{PO}\left(R_{PO} \parallel \frac{1}{g_{m2}}\right) \rightarrow \text{Moderate gain} \quad \therefore \text{ Can split } p_E : p_O \text{ with cross-amp } C_S$$

Shut with high v_{BO} $\rightarrow$ Isolated native (low-V_{TN0}) NMOS follower

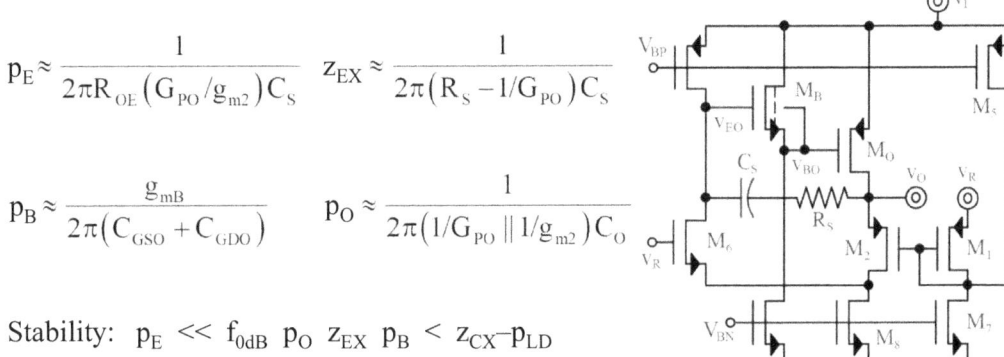

$$p_E \approx \frac{1}{2\pi R_{OE}\left(G_{PO}/g_{m2}\right)C_S} \quad z_{EX} \approx \frac{1}{2\pi\left(R_S - 1/G_{PO}\right)C_S}$$

$$p_B \approx \frac{g_{mB}}{2\pi\left(C_{GSO} + C_{GDO}\right)} \quad p_O \approx \frac{1}{2\pi\left(1/G_{PO} \parallel 1/g_{m2}\right)C_O}$$

Stability: $p_E \ll f_{0dB}$ p_O z_{EX} p_B $< z_{CX} - p_{LD}$

Design: v_R should supply current $\quad \beta_{FB} = 1$ (always) $\quad$ On-chip C_O with low Δi_{LD}

Digital LDO

Operation: Inverting feedback

Inverts after t_{CLK} delay $\quad\Bigg] + FB$

v_I limits v_{EO} swing

$\therefore$ Cycle-to-cycle Δv_{EO} gain $= 1$

$\rightarrow$ 1-LSB Oscillator about v_R $\rightarrow$ $\Delta d_{ADC} = 1$, $t_{OSC} = 2t_D = 2t_{CLK}$

Composition: Comparator (1-Bit ADC) + Counter = N-Bit ADC $\quad M_O\text{'s} = DAC$

Parameters: $R_{PO} = R_{CH}$ $\rightarrow$ Low PSR $\rightarrow$ Good for on-chip DSPs

$$A_{ADC0} = \frac{d_{ADC}}{\Delta v_{ID(LSB)}} = d_{ADC}\left(\frac{A_E}{v_I}\right) \quad \frac{1}{p_O} < t_D \quad A_{DAC0} = \frac{i_{PO(FS)}}{d_{ADC(FS)}} = i_{PO(LSB)} = \frac{v_I - v_O}{R_{CH(LSB)}}$$

$$i_{PO} = d_{ADC} i_{PO(LSB)} \quad\quad R_{PO} \approx \frac{R_{CH(LSB)}}{d_{ADC}} \quad\quad \Delta v_O \approx \Delta d_{ADC} i_{PO(LSB)}(R_{PO} \parallel R_{LD})$$

Chapter 7. Comparators

7.1. Introduction

7.2. Open Loop

7.3. Hysteretic

7.4. Regenerative

7.5. High Speed

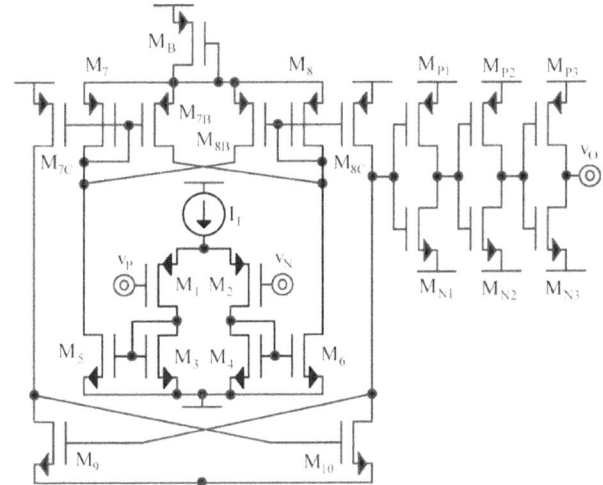

7.1. Introduction

Purpose: Compare analog inputs   1-bit analog–digital converter

"Trip" v_O when inputs crisscross   Polarity detector

Voltage Comparator:

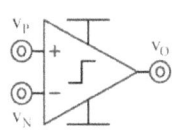

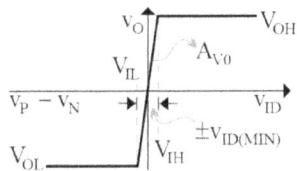

Applications:

Monitor voltages & currents   For protection, control, etc.

Parameters: Resolution $\rightarrow$ $\pm v_{ID(MIN)}$ $\rightarrow$ $f(A_{V0})$

Input Offset V_{OS} v_{IC} Range R_{ID}

Input Referred Noise v_{id}^* v_O Range   Delay t_P

Supply Noise $v_{dd/ss} \rightarrow f(A_{DD/SS})$ T_J Range   Slew Rate SR

Power $P_{DD/SS}$ $v_{DD} - v_{SS}$ Range   ...

105

A. Delay

Propagation Delay:

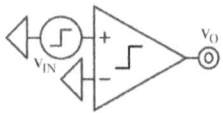

From $\quad V_{ID(MID)} = 0.5(V_{IL} + V_{IH})$

To $\quad V_{O(MID)} = 0.5(V_{OL} + V_{OH})$

Step Response

Possible Definitions:

$$t_P \equiv \text{Avg} \; \{ \; t_{P(RISE)} \, , \, t_{P(FALL)} \; \}$$

$$t_P \equiv \text{Max} \; \{ \; t_{P(RISE)} \, , \, t_{P(FALL)} \; \}$$

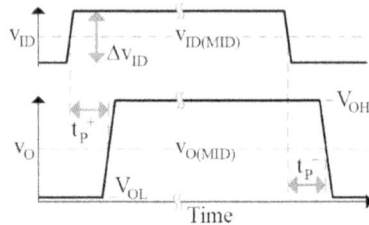

Bandwidth:

Small signals $\rightarrow$ Linear response $\rightarrow$ Poles delay small signals

$$A_V = \frac{A_{V0}}{1 + s/2\pi p_{BW}} = \frac{A_{V0}}{1 + \tau_{BW} s} \qquad \rightarrow \qquad v_O = \underbrace{v_{ID} A_{V0} \left[1 - \exp\left(\frac{-t}{\tau_{BW}} \right) \right]}_{v_{O(F)}}$$

Bandwidth Delay: Defined with step response $\rightarrow$ Instant $\Delta v_{ID} \geq \Delta v_{ID(MIN)}$

$$\Delta v_{O(MID)} = \frac{\Delta v_{O(MAX)}}{2} = \frac{V_{OH} - V_{OL}}{2} = \Delta v_{ID} A_{V0} \left[1 - \exp\left(\frac{-t_{P(BW)}}{\tau_{BW}} \right) \right]$$

Minimum drive: $\quad \Delta v_{ID(MIN)} \equiv \dfrac{V_{OH} - V_{OL}}{A_{V0}} \qquad \therefore \qquad \Delta v_{ID} = K_O \Delta v_{ID(MIN)}$

Overdrive factor: $\quad K_O \equiv \dfrac{\Delta v_{ID}}{\Delta v_{ID(MIN)}} \geq 1 \qquad t_{P(BW)} = \tau_{BW} \ln\left(1 - \dfrac{1}{2K_O} \right)^{-1} \leq 70\% \tau_{BW}$

Response: When $t_P \approx t_{P(BW)}$, negative exponential $\rightarrow$ Initially fast, then slow

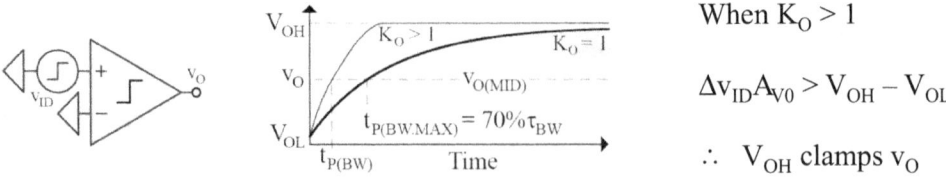

When $K_O > 1$

$$\Delta v_{ID} A_{V0} > V_{OH} - V_{OL}$$

$\therefore \quad V_{OH}$ clamps v_O

Tip: $\quad t = \tau_{BW} = \dfrac{1}{2\pi p_{BW}}$ when $v_O = 63\% v_{O(F)} \rightarrow$ Use to determine effective p_{BW}

106

SR Delay: Intentional C's & load C's delay large signals

$$SR \equiv \frac{dv_O}{dt} \quad \therefore \quad t_{P(SR)} = \frac{\Delta v_O}{SR} = \left(\frac{\Delta v_{O(MAX)}}{2}\right)\left(\frac{1}{SR}\right) = \left(\frac{V_{OH} - V_{OL}}{2}\right)\left(\frac{1}{SR}\right)$$

Total Delay: Low Δv_{ID} → Generates small signals that grow into large signals

$$\therefore \quad t_P \approx t_{P(BW)} + t_{P(SR)}$$

Example: Determine t_P when Δv_{ID} = 50 mV, A_{V0} = 100 V/V, V_{OL} = 0.5 V,

V_{OH} = 1.5 V, p_{BW} = 10 kHz, SR = 1 V/μs.

Solution: $\Delta v_{ID(MIN)} = \dfrac{V_{OH} - V_{OL}}{A_{V0}} = 10 \text{ mV} \quad \rightarrow \quad K_O = \dfrac{\Delta v_{ID}}{v_{ID(MIN)}} = 5$

$$t_{P(BW)} = \left(\frac{1}{2\pi p_{BW}}\right)\ln\left(1 - \frac{1}{2K_O}\right)^{-1} = 1.7 \text{ μs}$$

$$t_{P(SR)} = \left(\frac{V_{OH} - V_{OL}}{2}\right)\left(\frac{1}{SR}\right) = 500 \text{ ns}$$

$$t_P \approx t_{P(BW)} + t_{P(SR)} = 2.3 \text{ μs}$$

B. Noise

$v_{id}{}^*$ produces

uncertainty in transition:

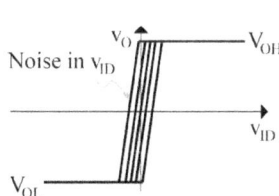

Hysteresis in trip points:

Rising v_{TH+} > Falling v_{TH-}

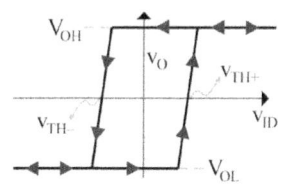

$v_{HYS} \equiv v_{TH+} - v_{TH-}$

$v_{id}{}^*$ produces "jitter":

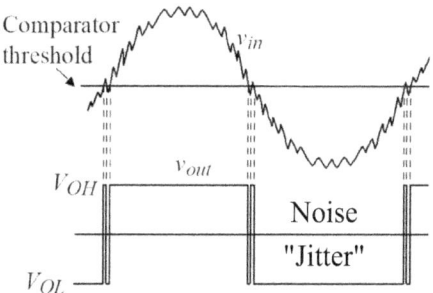

Sufficient v_{HYS} removes jitter:

$v_{HYS} > v_{id}{}^*$

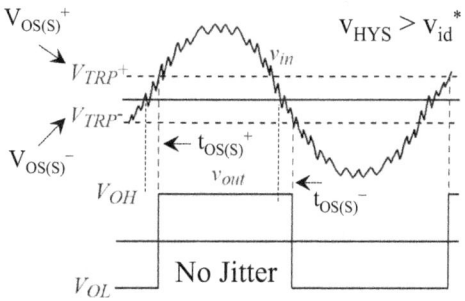

Hysteresis offsets/shifts transitions.

7.2. Open Loop: A. Mirror-Folded Stage

Behavior: Near transition → Comparators bias & amplify like amplifiers

Performance: Similar to mirror-folded amplifier → Analyze near transition

R_{ID}	A_{V0}
v_{IC} Range	p_O
v_{id}^*	p_F
A_{DD}	p_M
A_{SS}	z_{MX}
V_{OS}	SR

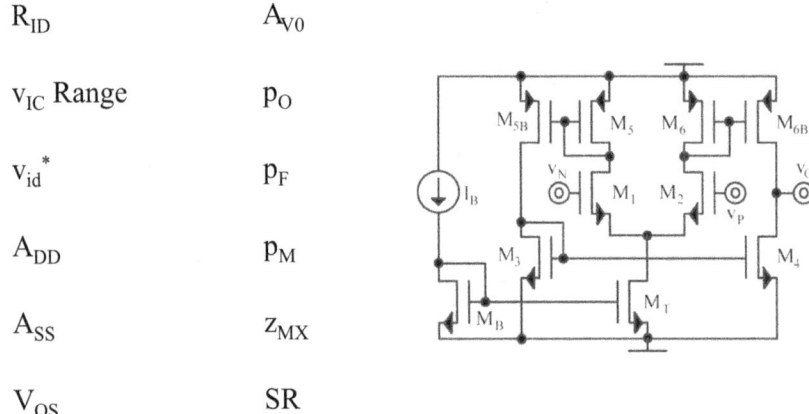

v_O Range: $V_{OH} = v_{DD}$ because $i_4 = 0$ $V_{OL} = v_{SS}$ because $i_{6B} = 0$

B. Class-A Output: i. CB/G Diff. Pair with CE/S Output

$v_{IC} \geq v_{SS} + v_{DS34(SAT)} + |v_{TP1A}| + v_{SD1A(SAT)}$

$V_{OH} = v_P - i_{B2}R_{CHA}$

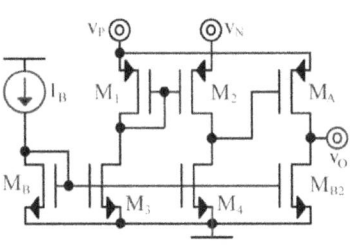

$R_{CHA} \approx \dfrac{1}{K_P'(W/L)_A \left(v_P - v_{SS} - |v_{TPA}| \right)}$

$A_{V0} \approx (-g_{m2})R_{O1}(-g_{mA})R_O$

$V_{OL} = v_{SS}$ because $i_A = 0$

$\left.\begin{array}{l} \text{If } v_{SG1} \approx v_{SGA} \\ \text{at trip point} \end{array}\right]$ $V_{OS} \approx 0 \pm \sqrt{\left(\dfrac{\Delta i_{12}^*}{g_{m12}}\right)^2 + \left(\dfrac{\Delta i_{34}^*}{g_{m12}}\right)^2}$

$p_{O1} \approx \dfrac{1}{2\pi R_{O1}(C_{GSA} + g_{mA}R_O C_{GDA})}$

$p_O \approx \left(\dfrac{g_{mA}C_{GDA}}{C_{GDA} + C_{GSA}}\right)\left(\dfrac{1}{2\pi C_{LD}}\right)$

$R_{IN1} = \dfrac{1}{g_{m1}} + r_{ds3} \approx r_{ds3}$ $R_{IN2} = \dfrac{r_{ds2} + r_{ds4}}{1 + g_{m2}r_{ds2}} \approx \dfrac{2}{g_{m2}}$

$v_{id}^* \approx \sqrt{2\left(\dfrac{i_{12}^*}{g_{m12}}\right)^2 + 2\left(\dfrac{i_{34}^*}{g_{m12}}\right)^2}$

$R_{O1} = R_{D2} \parallel r_{ds4}$ $R_O = R_{DA} \parallel r_{dsB2}$

ii. CE/S Diff. Pair with CE/S Output: Classic Class-A Structure

Performance: Similar to class-A op amp (near transition) without stabilizer

R_{ID} $\quad$ A_{V0}

v_{IC} Range $\quad$ p_{O1}

v_{id}^* $\quad$ p_O

A_{DD} $\quad$ p_M

A_{SS} $\quad$ z_{MX}

V_{OS}

$$R_{CHA} \approx \frac{1}{K_P'(W/L)_A\left(v_{SGA(MAX)} - |V_{TP0}|\right)}$$

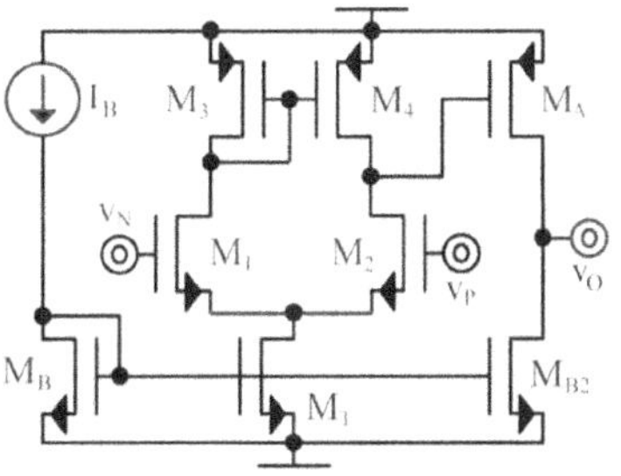

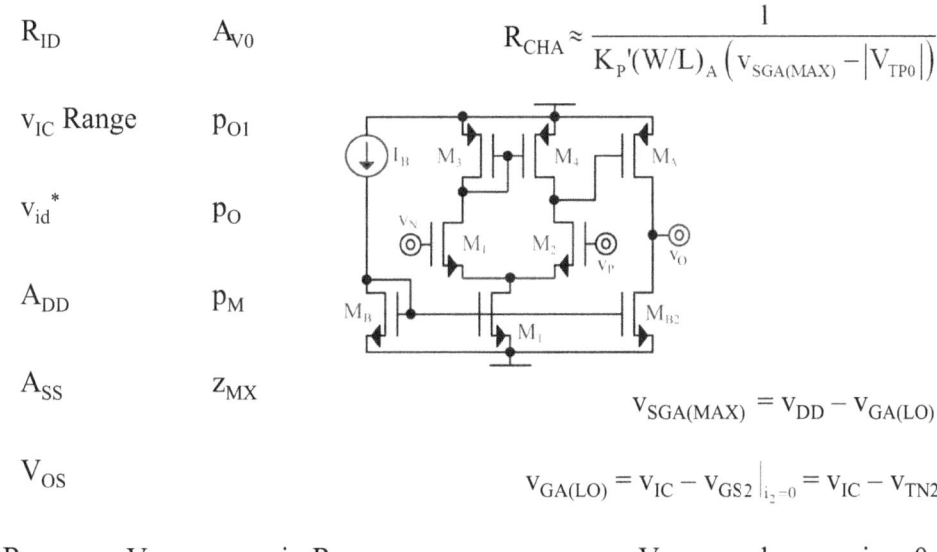

$$v_{SGA(MAX)} = v_{DD} - v_{GA(LO)}$$

$$v_{GA(LO)} = v_{IC} - v_{GS2}\Big|_{i_2=0} = v_{IC} - v_{TN2}$$

v_O Range: $\quad V_{OH} = v_{DD} - i_{B2}R_{CHA}$ $\qquad$ $V_{OL} = v_{SS}$ because $i_A = 0$

Delay

Trip Points:

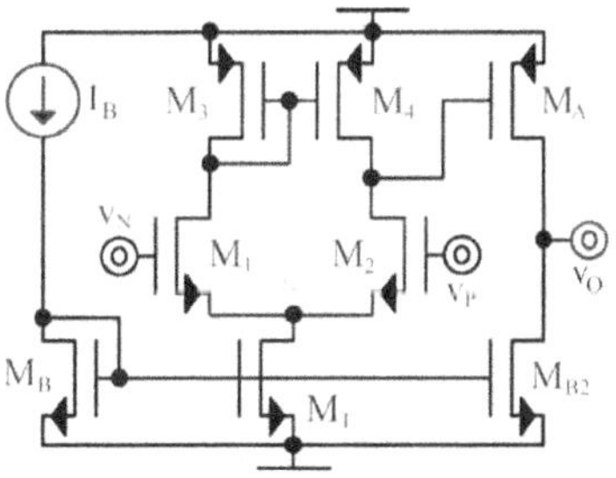

M_{1234} trips when v_P crosses v_N

M_A trips when v_{GA} crosses V_{THA}:

$$V_{THA} = v_{DD} - V_{SGA}\Big|_{i_{B2}} = v_{DD} - |V_{TP0}| - \sqrt{\frac{2i_{B2}}{K_P'(W/L)_A\left[1+\lambda_P(0.5)(v_{DD}-v_{SS})\right]}}$$

Delays:

i_T slews C_{GA} → $t_{GA} \propto \Delta v_{GA}$

$i_{A(MAX)}$ & i_{B2} slew C_O → $t_O \propto \Delta v_O$

Usually, $i_{A(MAX)} >> i_{B2}$ $\therefore$ $t_O^+ << t_O^-$

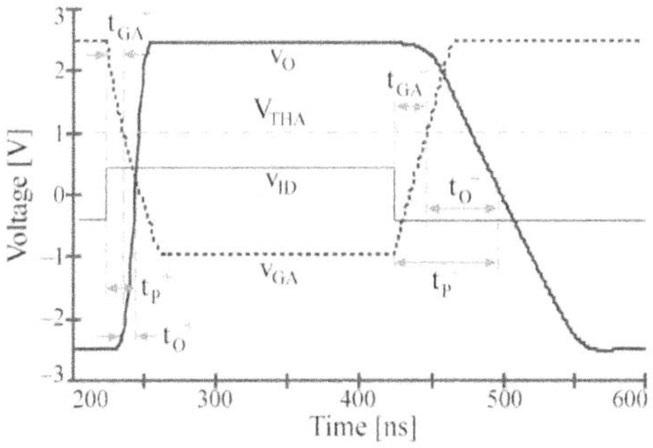

Propagation Delay:

Design so $t_{GA}{}^+ \approx t_{GA}{}^-$: $\quad V_{SG3}\big|_{0.5i_T} + V_{SGD}\big|_{0.5i_T} \equiv 2V_{SGA}\big|_{i_{B2}}$

$$t_P = t_{GA} + t_O \approx \left(\frac{C_{GA}}{i_T}\right)\Delta v_{GA} + \left(\frac{C_O}{i_{O(MAX\,MIN)}}\right)\left(\frac{v_{DD} - v_{SS}}{2}\right) \qquad \sqrt{\frac{0.5i_T}{(W/L)_{3D}}} \equiv \sqrt{\frac{i_{B2}}{(W/L)_A}}$$

$$\Delta v_{GA}{}^+ = v_{DD} - V_{THA} \quad < \quad \Delta v_{GA}{}^- = V_{THA} - v_{GA(LO)} \qquad \therefore \qquad \text{Limit } v_{GA(LO)} \text{ with } M_D$$

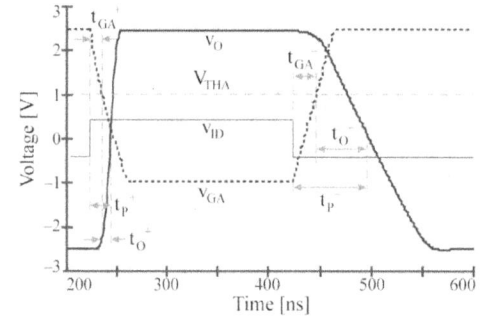 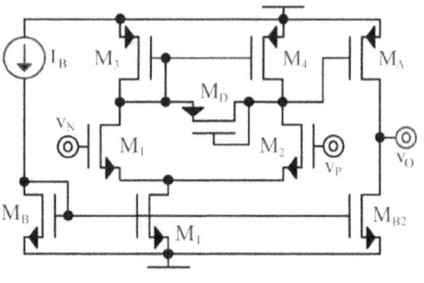

$$i_{O(MAX)} = i_{A(MAX)} - i_{B2} = f(v_{SGA(MAX)}) \qquad \rightarrow \qquad v_{GA(LO)} \text{ limits } t_O{}^+$$

$$i_{O(MIN)} = i_{B2} < i_{O(MAX)} \qquad \text{(usually)} \qquad \rightarrow \qquad i_{B2} \text{ limits } t_O{}^- \quad \therefore \quad t_O{}^+ < t_O{}^-$$

C. CMOS Inverter

Inverter compares v_I with threshold v_T

Operation: $\quad v_I < V_{TN0} + v_{SS} \qquad \rightarrow \qquad M_N = \text{Off} \qquad M_P = \text{On} \quad \therefore \quad M_P \text{ pulls } v_O \text{ high}$

$\qquad\qquad\quad v_I > v_{DD} - |V_{TP0}| \qquad \rightarrow \qquad M_P = \text{Off} \qquad M_N = \text{On} \quad \therefore \quad M_N \text{ pulls } v_O \text{ low}$

Threshold: $\quad$ MOSFETs balance when $v_O = 0.5(v_{DD} - v_{SS})$ at v_T

$\qquad\qquad$ Maximum noise immunity when $v_T = 0.5(v_{DD} - v_{SS})$

$\qquad\qquad$ Weaker MOSFET sets $i_I = i_N = i_P \quad \therefore \quad i_I$ peaks at v_T

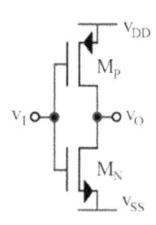

Min. Size: $\quad$ Short $t_I \quad \rightarrow \quad$ Low C's, R's $\quad \therefore \quad$ Use L_{MIN}'s, W_{MIN} for stronger K' device

Raise weaker W_{CH} until $i_P = i_N$ at $|v_{GS}| \equiv |v_{DS}| \equiv v_T$:

$$\frac{i_P}{i_N}\bigg|_{v_T}^{v_{GS} > V_{T9}} \approx \frac{W_P L_N K_P'\left(v_T - |V_{TP0}|\right)^2\left(1 + v_T\lambda_P\right)}{W_N L_P K_N'\left(v_T - V_{TN0}\right)^2\left(1 + v_T\lambda_N\right)} \equiv 1$$

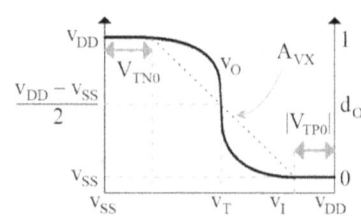

7.3. Hysteretic: A. Cross-Coupled Mirrors

Mechanics:

M_5's & M_6's close + feedback loop

$A_{LG+} < 1$ increases R_{O1} & R_{O2}

$A_{LG+} > 1$ establishes hysteresis

Rising Transition:

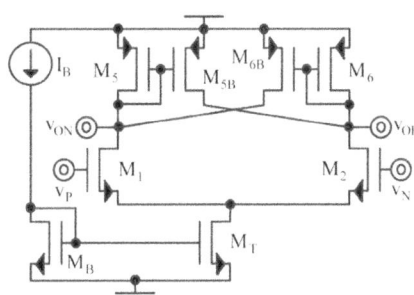

v_{OP} rises $\therefore$ v_{ON} starts high $\therefore$ M_5's = Off

$\quad\quad\quad\quad\quad\quad\quad\quad\quad\quad\quad\quad\quad\quad\quad\quad\quad\quad\quad$ $\times \equiv$ Crosses

v_{ON} falls when i_1 overcomes $i_{6B} \approx i_2(S_{6B}/S_6) \geq i_2$ $\therefore$ $A_{I(MIRROR)} \geq 1$

$\quad\quad\quad\quad\quad\quad\quad\quad\quad\quad\quad\quad\quad\quad\quad\quad\quad\quad$ v_{TN}'s cancel

$$V_{TH}{}^+ = v_{GS1} - v_{GS2} \approx \sqrt{\frac{2i_2(S_{6B}/S_6)}{K_N'S_1}} - \sqrt{\frac{2i_2}{K_N'S_2}} \geq 0$$

$\quad\quad\quad\quad\quad\quad\quad\quad\quad\quad\quad\quad\quad\quad\quad$ $i_1 + i_2 = i_2(S_{6B}/S_6) + i_2 = i_T$

Trip Point: $\quad i_1 = i_2(S_{6B}/S_6) \geq i_2 \quad \therefore \quad v_P \times v_N + V_{TH}{}^+ \quad \rightarrow \quad v_N \times v_P - V_{TH}{}^+$

Falling Transition:

v_{OP} falls $\therefore$ v_{OP} starts high $\therefore$ M_6's = Off

v_{OP} falls when i_2 overcomes $i_{5B} \approx i_1(S_{5B}/S_5) \geq i_1$

$$V_{TH}{}^- = v_{GS1} - v_{GS2} \approx \sqrt{\frac{2i_1}{K_N'S_1}} - \sqrt{\frac{2i_1(S_{5B}/S_5)}{K_N'S_2}} \leq 0$$

$\quad\quad\quad\quad\quad\quad\quad\quad\quad\quad\quad\quad\quad\quad\quad\quad$ v_{TN}'s cancel

$\quad\quad\quad\quad\quad\quad\quad\quad\quad\quad\quad\quad\quad\quad\quad$ $i_1 + i_2 = i_1 + i_1(S_{5B}/S_5) = i_T$

Trip Point: $\quad i_2 = i_1(S_{5B}/S_5) \geq i_1 \quad \therefore \quad v_N \times v_P + |V_{TH}{}^-| \quad \rightarrow \quad v_P \times v_N - |V_{TH}{}^-|$

Notes: Mirror translations set V_{TH} symmetry

$$A_{LG+} \approx (-g_{m5B})\left(\frac{1}{g_{m6}}\right)(-g_{m6B})\left(\frac{1}{g_{m5}}\right) = \left(\frac{S_{5B}}{S_5}\right)\left(\frac{S_{6B}}{S_6}\right) > 1 \quad \leftarrow \quad \text{For hysteresis}$$

Mirror-Folded Example

Motivation: Extend V_{OL} to v_{SS}

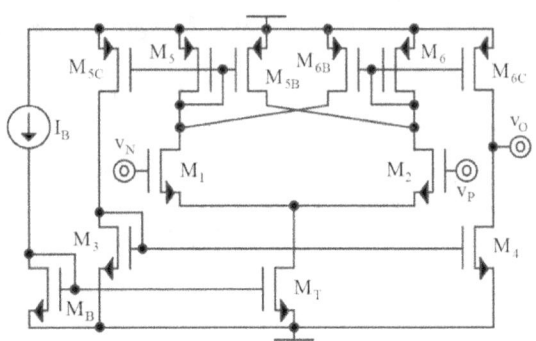

Given: $i_T = 20\ \mu A$, $S_{1256} = 2$,

$S_{5B} = 5$, $S_{6B} = 20$, $K_N' = 200\ \mu A/V^2$.

Solution:

V_{TH}^+: $i_2 + i_1 = i_1(S_{5B}/S_5) + i_1 = i_T$ when $i_1 = 5.7\ \mu A$

$$V_{TH}^+ = v_{GS2} - v_{GS1} \approx \sqrt{\frac{2i_1\left(S_{5B}/S_5\right)}{K_N'S_2}} - \sqrt{\frac{2i_1}{K_N'S_1}} = +98\ mV$$

Asymmetric

Hysteresis

V_{TH}^-: $i_1 + i_2 = i_2(S_{6B}/S_6) + i_2 = i_T$ when $i_2 = 1.8\ \mu A$

$$V_{TH}^- = v_{GS2} - v_{GS1} \approx \sqrt{\frac{2i_2}{K_N'S_2}} - \sqrt{\frac{2i_2\left(S_{6B}/S_6\right)}{K_N'S_1}} = -200\ mV$$

B. Offset Current

Operation: M_{FB} closes + feedback loop

i_H produces offset Δi_{12} that v_{ID} must overcome

Rising Transition: v_P & v_O start low

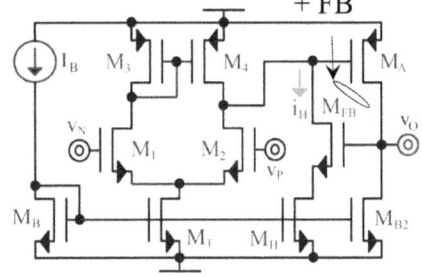

∴ M_{FB} = Off → $i_H = 0$

v_O rises when i_2 overcomes $i_4 \approx i_1$

∴ $V_{TH}^+ = 0$

Falling Transition: v_P & v_O start high   ∴ M_{FB} = On → $i_H > 0$

v_O falls when $i_4 \approx i_1$ overcomes $i_2 + i_H$ When $i_H \ll 0.5i_T$ $i_2 + i_H$

∴ $V_{TH}^- = v_{GS2} - v_{GS1} \approx \sqrt{\frac{2i_2}{K_N'S_2}} - \sqrt{\frac{2\left(i_2 + i_H\right)}{K_N'S_1}} \approx -\frac{i_H}{g_{m12}} < 0$ $i_1 + i_2 = i_T$

C. Offset Voltage

Operation:

Switches close + feedback loop

$\left[\begin{array}{l} v_H \text{ into } M_{H1} : M_{H2} \text{ produces offset } \Delta i_H \\ \text{Switches steer } \Delta i_H \text{ to reinforce state} \end{array}\right.$

v_H produces offset Δi_H that v_{ID} must overcome

Adds v_{OS} when v_O = Low $\qquad \therefore \; v_P^+ \times v_N + v_{OS} \; \rightarrow \; V_{TH}^+ = +v_{OS}$

Subtracts v_{OS} when v_O = High $\; \therefore \; v_P^- \times v_N - v_{OS} \; \rightarrow \; V_{TH}^- = -v_{OS}$

$$v_{OS} \approx v_H \left(\frac{g_{mH12}}{g_{m12}} \right)$$

Design

Requirement:

$v_H < v_{ID(MAX)}$

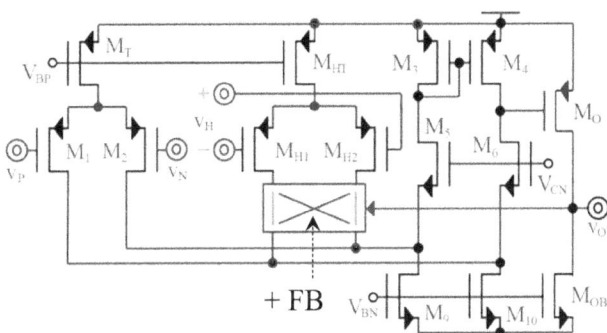

+ FB

D. Hysteretic CMOS Inverter

Operation: v_O rises when $i_{P1} > i_{PT}$ $\;\rightarrow\;$ $i_{P1} \equiv i_{PT}$ at v_{T-}

M_{P2} starts conducting when $v_P = v_{T-} + v_{TP2}$

$$\frac{i_{P1}}{i_{PT}} \approx \frac{W_{P1}L_{PT}\left(v_{DD} - v_{T-} - |V_{TP0}|\right)^2}{W_{PT}L_{P1}\left(v_{T-} + |v_{TP2}| - |v_{TPT}|\right)^2} \equiv 1$$

v_O falls when $i_{N1} > i_{NT}$ $\;\rightarrow\;$ $i_{N1} \equiv i_{NT}$ at v_{T+}

M_{N2} starts conducting when $v_N = v_{T+} - v_{TN2}$

$$\frac{i_{N1}}{i_{NT}} \approx \frac{W_{N1}L_{NT}\left(v_{T+} - V_{TN0}\right)^2}{W_{NT}L_{N1}\left[v_{DD} - (v_{T+} - v_{TN2}) - v_{TNT}\right]^2} \equiv 1$$

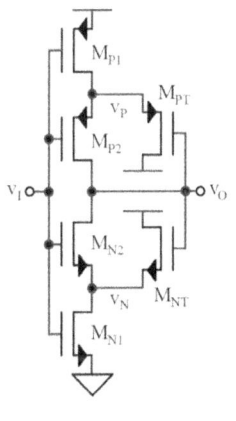

Design: Design inverter $W_{N/P}$, adjust W_{N12} & W_{P12}, set $v_{T+/-}$ with W_{PT} & W_{NT}.

M_{N12} & M_{P12} double R's to supplies $\quad \therefore \; W_{N12} \equiv 2W_N, \; W_{P12} \equiv 2W_P$

7.4. Regenerative: A. Mechanics

+ Feedback: Latches v_O's to $V_{OH/OL}$ when $A_{LG+} > 1$

                     Regenerates v_O's $\therefore$ Accelerates transition & expands swing

Location: At the input $\rightarrow$ Shifts trip points to establish hysteresis

               At the output $\rightarrow$ Regenerates response

                                    A_{V1} suppresses effects on trip points

Latch: CE/S Diff. Pair   Operation:

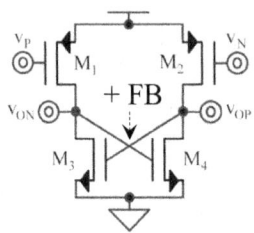

                                    Small v_{ID} produces imbalance

                                    + FB regenerates & grows

                                    initial imbalance

              Cross-Coupled Pair   until v_O's latch

B. Delay

Small v_{ID}:

Diff. pair: $i_1 = 0.5v_{id}g_{m1} \approx -i_2 = -(-0.5v_{id}g_{m2})$ $\therefore$ $\Delta v_{ON(INI)} = -\Delta v_{OP(INI)}$

Cross pair: $R_{EQ} \equiv \dfrac{\Delta v_{ON}}{\Delta i_3} \approx \dfrac{\Delta v_{ON}}{\Delta v_{OP}g_{m3}} = -\dfrac{1}{g_{m3}}$ $C_{EQ} \approx C_{GS}$ $\rightarrow$ – Time Constant τ_L

$\therefore$ $\Delta v_{O(FIN)} = \Delta v_{O(INI)} \exp\left(\dfrac{-t}{\tau_L}\right) \approx \Delta v_{O(INI)} \exp\left(\dfrac{g_{m3}t}{C_{GS}}\right)$ $\rightarrow$ + Exponential

                                                  (Right-Half-Plane Pole)

Propagation Delay:

$$t_P = \left(\frac{C_{GS}}{g_{m3}}\right) \ln\left(\frac{0.5\Delta v_{O(FIN)}}{\Delta v_{O(INI)}}\right) = -\tau_L \ln\left(\frac{1}{2K_D}\right)$$

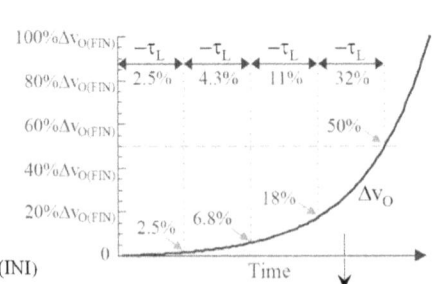

     $\rightarrow$ Slow with low $\Delta v_{O(INI)}$

     $\rightarrow$ Exponentially faster with higher $\Delta v_{O(INI)}$

     $\rightarrow$ $K_D \equiv$ Initial Drive Fraction ≤ 1

                                           Initially slow, then fast

7.5. High Speed

Linear Response

Regenerative Response

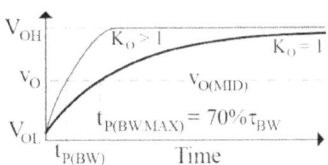

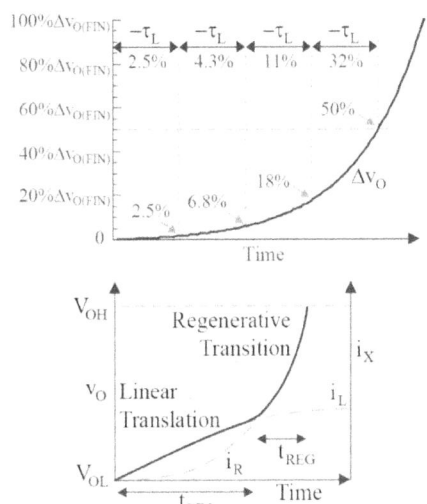

Linear Response:

  Initially fast, then slow

Regenerative Response:

  Initially slow, then fast

Fast Overall Response:

  Linear input amplifies low Δv_{ID} drive $\rightarrow$ Last linear Δv_O grows

  Regenerative output accelerates last linear Δv_O

A. v_{ID} Amplifier

Linear stages with regenerative output:

$$A_{V0} \geq \frac{V_{OH} - V_{OL}}{\Delta v_{ID(MIN)}} = A_{VL0}A_{VR0} = A_{VL0}A_{GR0}(R_L \parallel R_R)$$

A_{VL0} suppresses hysteresis produced by regeneration $\rightarrow$ $A_{VL0} \geq 10$

Gain per stage $A_{VX0} \propto R_X$

$t_{P(BW)} \propto$ Pole time constant $\tau_X \propto R_X$

$\left.\begin{array}{l}\\\\\end{array}\right]$ Limit R_X $\rightarrow$ Limit A_{VX0}

$\downarrow$

Gain per stage

$t_{P(SR)} \propto$ Swing Δv_X $\rightarrow$ Limit Δv_X

Accelerate ("regenerate") last output $\rightarrow$ $A_{RLG+} > 1$

B. i_O Driver

Drive large C_{LD} → Supply/pull substantial i_{LD}

Inverter: M_P charges C_O when v_{IN} = Low

M_N discharges C_O when v_{IN} = High

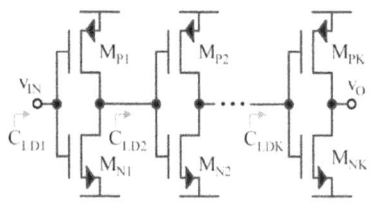

Balance t_P's: Center threshold V_{TH} between supplies

∴ Balance MOS i_D strengths

→ Use W's that balance K_N' : K_P' and $|v_{TN}|$: $|v_{TP}|$

Shorten t_P's: Lighten inter-stage loads C_{LD}'s

∴ C_{LD1} ≡ Minimum-size stage

Build drive

∴ Cascade increasingly larger stages

Lowest delay when inter-stage fan out $f_o \equiv \dfrac{S_{X+1}}{S_X} = 3.6$

C. Example

Amplifier

Low-Δv_{O34} M_{1234}:

$$A_{V10} \approx \frac{g_{m12}}{g_{m34}} \equiv 10$$

Low-Δv_{O78} M_{5678B}:

$$A_{V20} \approx \frac{g_{m56}\left(\dfrac{1}{g_{m78}}\right)}{1 - \left(\dfrac{S_{7B}}{S_7}\right)\left(\dfrac{S_{8B}}{S_8}\right)} \equiv 10 \qquad A_{78LG+} < 1$$

Reg. high-$\Delta v_{O8/10}$ latch $M_{78C9/10}$: $A_{9/10LG+} > 1$

$$A_{V30} \approx g_{m78C}\left(\frac{r_{ds}}{2} \,\|\, \frac{1}{g_{m9/10}}\right) \geq 10 \qquad M_B \text{ boosts } M_{78C}\text{'s gate drive}$$

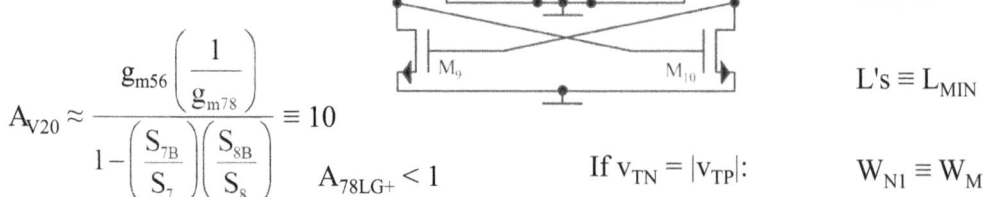

Driver

L's ≡ L_{MIN}

If $v_{TN} = |v_{TP}|$: $W_{N1} \equiv W_{MIN}$

$W_{PX} \equiv \left(\dfrac{K_N'}{K_P'}\right)W_{NX}$ $W_{N2} \equiv 4W_{N1}$

$W_{N3} \equiv 4W_{N2}$

116

Chapter 8. Reference Circuits

8.1. Voltage Primitives

8.2. PTAT Core

8.3. Temperature Independence

8.4. Current References

8.5. Voltage References

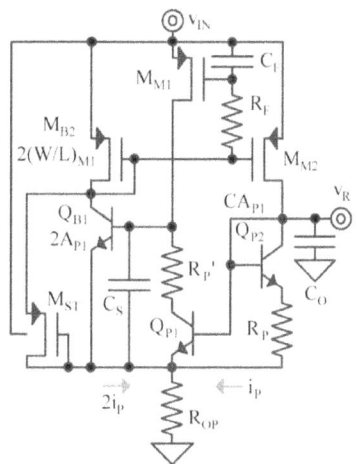

8.1. Voltage Primitives

Diode Voltage:

$$i_D = I_S\left[\exp\left(\frac{v_D}{V_t}\right) - 1\right] \qquad \rightarrow \qquad v_D = V_t \ln\left(\frac{i_D}{I_S} + 1\right) \approx \left(\frac{K_B T_J}{q_E}\right) \ln\left(\frac{i_D}{I_S}\right)$$

$v_D \approx 600\text{–}700 \text{ mV} \pm 2\%$ at T_{ROOM} $- 2.2 \text{ mV/°C}$ $\rightarrow$ Falls with T_J

Exponential across 4–5 decades of i_D

Logarithm suppresses effects of Δi_D on v_D

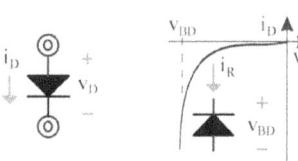

Breakdown Voltage:

$v_{BD(ZENER)} < 6 \text{ V} \pm 2\%\text{–}4\%$ at T_{ROOM} $\rightarrow$ $v_{BD(ZENER)}$ falls with T_J

$v_{BD(AVALANCHE)} > 4 \text{ V} \pm 2\%\text{–}4\%$ at T_{ROOM} $\rightarrow$ $v_{BD(AVALANCHE)}$ rises with T_J

Typical N^+P Junctions: $v_{BD} \approx 5\text{–}7 \text{ V} + 2$ to 4 mV/°C

Logarithmic-like relation suppresses effects of Δi_D on v_{BD}

117

Sub-v_T Gate–Source Voltage:

$$i_D\Big|_{v_{DS}>3V_t}^{0<v_{GS}<v_T} \approx \left(\frac{W}{L}\right)I_S \exp\left(\frac{v_{GS}}{n_I V_t}\right)\exp\left(\frac{-v_T}{n_I V_t}\right) \qquad \rightarrow \qquad v_{GS} \approx n_I V_t \ln\left[\frac{i_{D(SUB)}}{(W/L)I_S'}\right]$$

$v_{GS} < v_T$ 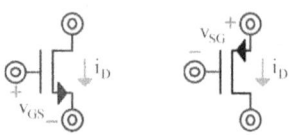 Where $n_I = 1\text{--}2$

Exponential (in sub-v_T) across 1–2 decades of i_D

Logarithm suppresses effects of Δi_D on v_{GS}

Inverted Gate–Source Voltage:

$$i_D\Big|_{v_{DS}>v_{GST}}^{v_{GS}>v_T} \approx 0.5\left(\frac{W}{L}\right)K_{N\,P}'\left(v_{GS}-v_T\right)^2 \quad \rightarrow \quad v_{GS} = v_T + v_{DS(SAT)} \approx v_T + \sqrt{\frac{2i_{D(INV)}}{K_{N/P}'(W/L)}}$$

$\Delta v_T \approx \pm 50$ mV $\qquad\qquad \Delta K_{N/P}' \approx \pm 20\%$ $\qquad\qquad v_T$ & $K_{N/P}'$ fall with T_J

$\therefore\quad \Delta v_{GS} \approx \pm 5\%$ to $\pm 10\%$ at T_{ROOM} $\qquad \rightarrow \quad$ Higher variation than v_D

Quadratic (in inversion) across 1–2 decades of i_D

Square-root suppresses effects of Δi_D on v_{GS} $\quad \rightarrow \quad$ Lower suppression than ln

8.2. PTAT Core

Proportional to Absolute Temperature: $V_t = \dfrac{K_B T_J}{q_E} = v_{PTAT} \propto T_J$

V_t Features: Predictable, linear across T_J, consistent across operating conditions

Generation:

$$\Delta v_D \equiv v_{D1} - v_{D2} \approx V_t \ln\left(\frac{i_{D1}I_{S2}}{I_{S1}i_{D2}}\right) = V_t \ln\left(\frac{i_{D1}A_{J2}}{A_{J1}i_{D2}}\right) = v_{PTAT} \qquad I_S \propto \text{Area } A_J$$

$$\Delta v_{GS} \equiv v_{GS1} - v_{GS2}\Big|_{Sub\text{-}v_T} \approx n_I V_t \ln\left[\frac{i_{D1(SUB)}(W/L)_2}{(W/L)_1 i_{D2(SUB)}}\right] \equiv n_I V_t \ln\left(\frac{i_{D1(SUB)}S_2}{S_1 i_{D2(SUB)}}\right) = v_{PTAT}$$

ln term = Constant $\qquad\qquad$ when $\qquad\qquad$ i_D's, A_J's, (W/L)'s match

$i_{PTAT} = \dfrac{v_{PTAT}}{R_P} \propto T_J$ $\qquad\qquad$ when $\qquad\qquad$ R_P does not drift with T_J

On-chip poly R's vary $\pm 20\%$ $\qquad \rightarrow \qquad$ Adjust ("trim") R_P when required

A. Latched CB/G Cells

Composition:

Current mirror $M_{M1} : M_{M2}$ matches i_D's $\rightarrow$ $i_{P1} \approx i_{P2}$

CG pair $M_{B1} : M_{B2}$ matches v_S's $\rightarrow$ $v_{GSB1} \approx v_{GSB2}$ $\therefore$ $v_{SB1} \approx v_{SB2}$

PTAT pair $D_{P1} : D_{P2}$ produces v_{PTAT} $\rightarrow$ $v_R \approx v_{D1} - v_{D2} \approx V_t \ln\left(\dfrac{A_{P2}}{A_{P1}}\right) = V_t \ln C$

Positive feedback latches cell into stable state $\rightarrow$ $M_{B2} : M_{M2} : M_{M1} : M_{B1}$

Stable states:

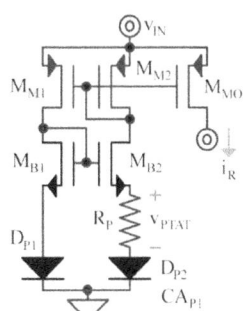

PTAT: $\quad i_R \approx i_{P1} \approx i_{P2} \approx \dfrac{\Delta v_{D12}}{R_P} \approx \left(\dfrac{V_t}{R_P}\right) \ln C$

Off: $\quad i_R \approx i_{P1} \approx i_{P2} = 0$

$\therefore$ $\quad$ Bi-stable $\rightarrow$ Requires starter

Voltage error: $\quad v_{SDM1} \neq v_{SDM2} \qquad v_{DSB1} \neq v_{DSB2}$

Cell Variations

Diode: Headroom: $v_{IN(MIN)} = \text{Max}\{v_D + v_{GS} + v_{SD(SAT)}, \; v_D + v_{DS(SAT)} + v_{SG}\}$

BJT: Headroom: $v_{IN(MIN)} = \text{Max}\{v_{BE} + v_{SD(SAT)}, \; v_{PTAT} + v_{CE(MIN)} + v_{SG}\}$

$\quad$ Base-current error: $\quad i_{M12} = i_{C2} = i_{C1} + i_{B1} + i_{B2}$ $\rightarrow$ $\quad i_{C1} < i_{C2}$

MOS: Headroom: $v_{IN(MIN)} = \text{Max}\{v_{GS} + v_{SD(SAT)}, \; v_{PTAT} + v_{DS(SAT)} + v_{SG}\}$

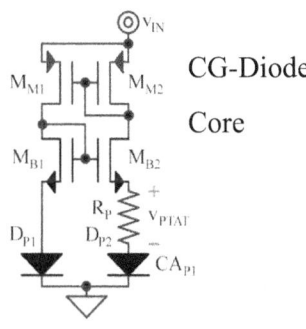
CG-Diode
Core

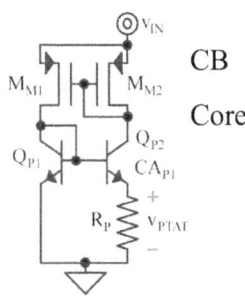
CB
Core

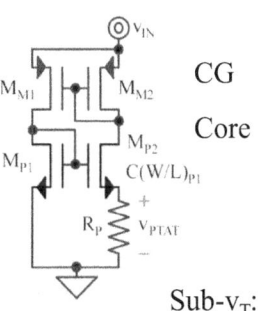
CG
Core

Sub-v_T:

$i_R \approx \dfrac{\Delta v_{D12}}{R_P} \approx \dfrac{V_t}{R_P}\ln\left(\dfrac{A_{P2}}{A_{P1}}\right)$ $\qquad$ $\dfrac{\Delta v_{BE12}}{R_P} \approx \dfrac{V_t}{R_P}\ln\left(\dfrac{A_{P2}}{A_{P1}}\right)$ $\qquad$ $\dfrac{\Delta v_{GS12}}{R_P} \approx \dfrac{n_1 V_t}{R_P}\ln\left(\dfrac{S_{P2}}{S_{P1}}\right)$

B. Starter

Bi-stable: Cell latches to $i_R = i_{PTAT}$ or $i_R = 0$

Fix: Ensure + FB current $i_{FB+} \neq 0$ $\rightarrow$ $i_R \neq 0$ $\therefore$ Latches to i_{PTAT}

i. Continuous

Concept: Feed or pull i_{ST} into/from + FB loop

i_{ST} = Error $\therefore$ Keep i_{ST} very low $\rightarrow$ Define with very high R_{ST}

Options: Narrow-W long-L poly-silicon R

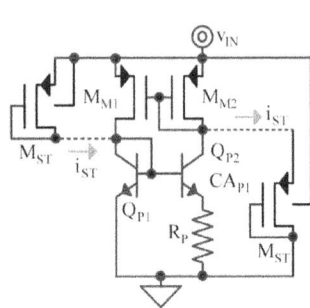

Narrow-W long-L JFET (diffusion R)

Narrow-W long-L high-v_T MOSFET

CMOS: Choose lowest $K_{N/P}'v_{GST}$ MOSFET type

Body effect can raise v_T

Example

Target: i_{ST} = 250 nA at $T_J = 27°$ when $n_I = 1.75$, $v_{SG} = 1.5$ V, $v_{BS} = 1$ V, $W_{MIN} = 1$ μm,

$V_{TP0} = -400$ mV, $K_p' = 40$ μA/V^2, $\gamma_P = 500$ mV√V, $2\psi_B = 600$ mV.

Solution:

$$\left|v_{TP}\right| = \left|V_{TP0}\right| + \gamma_P \left(\sqrt{2\psi_B - v_{SB}} - \sqrt{2\psi_B}\right) = 640 \text{ mV}$$

Low i_{ST} $\therefore$ Low S_{ST} $\rightarrow$ $W_{ST} = W_{MIN} = 1$ μm $S_{ST} < 1$

$$v_{SD(SAT)} \approx \sqrt{\frac{2i_{ST}}{K_p'S_{ST}}} > \sqrt{\frac{2i_{ST}}{K_p'(1)}} = 110 \text{ mV} > 2n_I V_t = 90 \text{ mV} \quad \therefore \text{ Inverted}$$

Diode-connected $\rightarrow$ $v_{SD} = v_{SG} \approx \left|v_{TP}\right| + v_{SD(SAT)} > v_{SD(SAT)}$ $\therefore$ Saturated

$$i_{ST} \approx \left(\frac{W_{ST}}{L_{ST}}\right)\left(\frac{K_p'}{2}\right)\left(v_{SG} - \left|v_{TP}\right|\right)^2 \quad \rightarrow \quad L_{ST} = 59 \text{ μm}$$

120

ii. On Demand: Voltage Mode

Concept: Feed i_{ST} when i_p nears 0 state $\rightarrow$ Sense when $v_{BE/GSP}$ drops

Diode Stack: Q_{S5} feeds i_{ST}

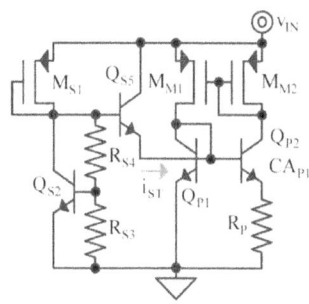

    Narrow-W long-L M_{S1} feeds Q_{S2}

    R_{S4} diode-connects Q_{S2}

    R_{S34} multiplies v_{BES2}:   Design Aim

$$v_{BS5} \approx v_{BES2}\left(\frac{R_{S3} + R_{S4}}{R_{S3}}\right) \equiv 1.5 v_{BES2}$$

    Q_{S5} Feeds:

BJT Sensor: If PTAT Cell = On: $v_{BES5} = 1.5 v_{BES2} - v_{BEP} \approx 0.5 v_{BE}$ $\rightarrow$ $i_{ST} \approx 0$

    If PTAT Cell = Off: $v_{BES5} \approx 1.5 v_{BES2} - 0 = 1.5 v_{BE}$ $\rightarrow$ $i_{ST} > 0$

CE/S Differential-Pair Sensor: Q_{S4} pulls i_{ST}

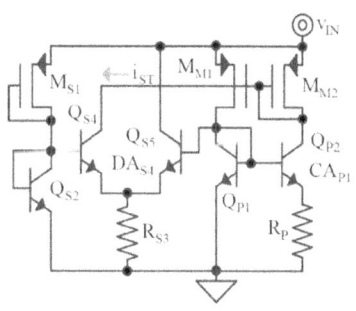

    Narrow-W long-L M_{S1} feeds Q_{S2}

    High R_{S3} biases diff. pair $Q_{S4} : Q_{S5}$

    $A_{S5} = DA_{S4} > A_{S4}$

$\therefore$ $Q_{S4} : Q_{S5}$ favors Q_{P1}'s v_{BEP} over Q_{S2}'s v_{BES2}:

    If PTAT Cell = On: $v_{BES2} \approx v_{BEP}$ $\therefore$ $i_{ST} \ll i_{S5}$

$\therefore$ $v_{BES4} \approx V_t \ln\left(\dfrac{i_{ST}}{I_{S4}}\right) \approx v_{BES5} \approx V_t \ln\left(\dfrac{i_{S5}}{DI_{S4}}\right)$ $\rightarrow$

    If PTAT Cell = Off: $v_{BES2} \gg v_{BEP} \approx 0$ $\rightarrow$

$$i_{ST} \approx \frac{i_{S5}}{D} \approx \frac{i_{S3}}{D+1}$$

$$\uparrow$$

$$i_{S3} \approx \frac{i_{S5}}{D} + i_{S5}$$

$$i_{ST} = \text{low, but} \neq 0$$

$$i_{ST} \approx i_{S3} \gg i_{S5}$$

Current Mode

Concept: Feed i_{ST} when i_P nears 0 state → Sense when i_P drops

i_P Comparator:

M_{MS} mirrors (senses) i_P

Narrow-W long-L M_{S1} feeds starter

M_{S2} compares i_P & i_{S1}:

∴ If PTAT Cell = On: $i_P \gg i_{S1}$ → $v_{ST} \approx v_{IN}$ ∴ M_{S2} = Off

If PTAT Cell = Off: $i_P < i_{S1}$ → $v_{ST} \approx 0$ ∴ $i_{ST} > 0$

C_{ST} deglitches i_{ST}: $\left. \begin{array}{l} M_{M2} \text{ couples } v_{in} \text{ noise to } v_{GM} \\ C_{ST} \text{ couples } v_{in} \text{ noise to } v_{ST} \end{array} \right\}$ M_{S2} rejects v_{in}

(E.g., $i_P \geq 10$ nA) → $C_{ST} \equiv$ Coupling common-mode capacitor to v_{IN}

Design Note: Ensure $i_P \gg i_{S1} \gg$ noise $i_n{}^*$ across T_J & fabrication corners

C. 1st-Order Corrections

Errors: Voltage & Base Current

Fix: Match $v_{CE/DS}$'s & i_B's

How: Alter diode-connecting loop

i–ii: $A_{G/V}$ diode-connects $M_{M2/P1}$ A_G matches mirror's v_{DS}'s

ii: A_V matches v_R's ∴ $i_1 = i_2$ R_P' matches v_{CE}'s when $R_P' \equiv R_P$

iii: $Q_{P1} : Q_{B3} : M_{B3}$ diode-connects M_{M2}

Aims: $i_{B3} = i_{B1} + i_{B2}$ ∴ $i_{C3} \equiv 2i_{C12}$ → $S_{M3} \equiv 2S_{M12}$

$$v_{BE3} = v_{BE1} \quad \therefore \quad \frac{i_{C1}}{A_{P1}} \equiv \frac{i_{C3}}{A_{B3}} \quad \rightarrow \quad A_{P3} \equiv 2A_{P1}$$

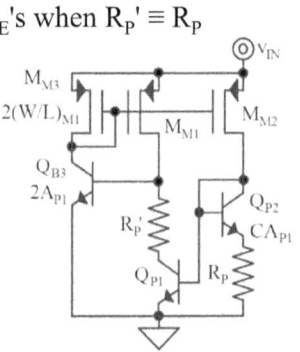

D. Stabilization

Loops: + FB latches i_P to i_{PTAT} (with starter)   − FB diode-connects mirror

Aims: $|A_{LG+}| > 1 > |A_{LG-}|$ when $i_P \approx 0$ E.g.:

$\therefore$ Use i_{ST} to raise A_{LG+}

A_{LG+} inverts A_{LG-} when $i_P \approx i_{PTAT}$

$\therefore$ Use RC filter to suppress A_{LG+}

Use R_P to: Degenerate A_{LG+}

or   Boost/favor A_{LG-}

Use common-mode C_S's to stabilize − FB loop(s)   Boosts A_{LG-}

Note: Multiple loops $\rightarrow$ Multiple stable states $\therefore$ Often difficult to start

Startup & stabilization can be empirical $\rightarrow$ Trial & error with simulations

8.3. Temperature Independence: A. CTAT

Complementary to Absolute Temperature: v_{CTAT} falls linearly with T_J

Candidates: $v_{D/BE}$ $|V_{TN/P0}|$ $K_{N/P}'$

Limitation: Fall with T_J is not perfectly linear

i_{CTAT}:

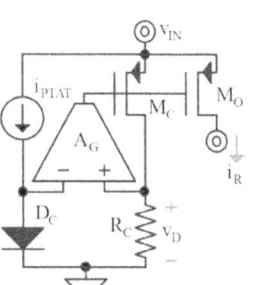

Voltage-

Matching

Amp A_G

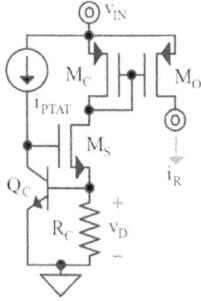

Current-

Sensing

Transistor M_S

A_G diode-connects M_C $\rightarrow$ $M_C : M_O$ Mirror

A_G series-mixes inputs $\rightarrow$ $v_D \approx v_{RC}$

$M_C : M_O$ mirrors $i_R \approx i_{RC} \approx \dfrac{v_D}{R_C} \approx i_{CTAT}$

M_S diode-connects Q_C $\rightarrow$ $v_{RC} = v_{BE}$

M_S series-samples i_{RC} $\rightarrow$ $i_S \approx i_{RC}$

$M_C : M_O$ mirrors $i_R \approx i_{RC} \approx \dfrac{v_{BE}}{R_C} \approx i_{CTAT}$

B. Compensation

Concepts: PTAT cancels CTAT 1st Order $\equiv$ Cancels T_J^1 terms

Diode Voltage: Taylor-series expansion of $v_D \approx V_t \ln(i_D/I_S)$ when $i_D = K_A T_J^X$

$$v_D \approx V_{BG} - (V_{BG} - v_{D(ROOM)})\left(\frac{T_J}{T_{ROOM}}\right) - (\eta - x)V_t \ln\left(\frac{T_J}{T_{ROOM}}\right) \qquad \text{Where } \eta \approx 4$$

$\therefore \ T_J^0 + T_J^1 + T_J \ln T_J$ Terms $\rightarrow$ v_D = Nonlinear with T_J $V_{BG} \equiv$ Bandgap Voltage

CTAT Primitive:

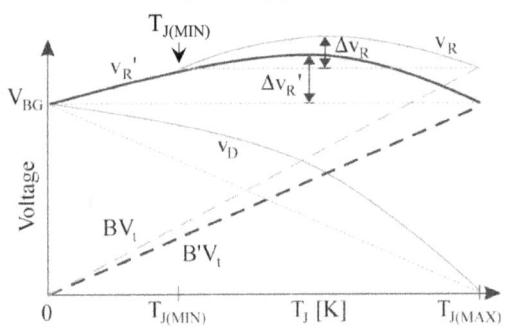

$v_D \approx V_{BG} \approx 1.2$ V at 0 K

$v_D \approx 600\text{–}700$ mV at T_{ROOM}

PTAT Primitive:

$V_t = 0$ V at 0 K $\qquad V_t \propto T_J$

1st-Order Bandgap v_R: $v_R' \approx V_{BG} + \Delta v_R'$

$T_{J(MIN)} \neq 0$ K $\ \therefore \ \Delta v_R' \neq$ Optimal $\rightarrow$ Add v_{PTAT} $\left[\begin{array}{l} \text{Optimal } \Delta v_R < \Delta v_R' \\ \text{Optimal } v_R > V_{BG} \end{array}\right.$

Bandgap Approximation: $v_R = v_D + v_{PTAT} \approx V_{BG} \approx 1.2$ V

$\rightarrow$ at T_{ROOM}: $v_{D(ROOM)} \approx 0.6\text{–}0.7$ V $\therefore$ $v_{PTAT(ROOM)} \approx 0.5\text{–}0.6$ V

1st-Order Bandgap i_R: Without T^1 Terms Nonlinearity with $-T_J$ drift

$$i_R = i_C + i_P = \frac{v_D}{R_C} + \frac{\Delta v_D}{R_P} = \frac{v_D}{R_C} + \frac{V_t \ln C}{R_P} \approx \frac{V_{BG}}{R_C} - \left[\frac{(\eta - x)V_t}{R_C}\right]\ln\left(\frac{T_J}{T_{ROOM}}\right) \approx \frac{V_{BG}}{R_C}$$

Cancel T^1 terms: $\dfrac{V_{t(ROOM)} \ln C}{R_P} \equiv \left(\dfrac{V_{BG} - v_{D(ROOM)}}{R_C}\right)\left(\dfrac{T_{ROOM}}{T_{ROOM}}\right)$

$\therefore$ $\dfrac{R_C}{R_P} \approx \dfrac{V_{BG} - v_{D(ROOM)}}{V_{t(ROOM)} \ln C}$

Nonlinearity: $T \ln T$ term $\rightarrow$ Nonlinear Δv_R curvature

If $i_D = i_{PTAT} = K_B T_J^X = K_B T_J^1$ $\ \therefore \ \eta - x = \eta - 1$ $\rightarrow$ Lower curvature

124

Example

Target: Determine R_C & R_P so $i_R = 5\ \mu A$ when $v_D = 650$ mV at T_{ROOM}, $C = 8$.

Solution:

$$i_R = \frac{v_D}{R_C} + \frac{V_t \ln C}{R_P}\bigg|_{T_J=0} = \frac{V_{BG}}{R_C} + 0 \approx \frac{1.2}{R_C} \equiv 5\ \mu A \quad \rightarrow \quad R_C = 240\ k\Omega$$

$$\frac{v_{PTAT(ROOM)}}{R_P} = \frac{V_{t(ROOM)} \ln C}{R_P} \equiv \frac{V_{BG} - v_{D(ROOM)}}{R_C} \quad \rightarrow \quad R_P = 23.6\ k\Omega$$

Error: Centered between 0 K and $T_{J(MAX)}$ Not between $T_{J(MIN)}$ and $T_{J(MAX)}$

$\therefore$ Simulate/adjust $R_{P/C}$ until curvature balances across $T_{J(MIN)} - T_{J(MAX)}$

8.4. Current References: A. Latched Core

Compact Solution: Integrate CTAT into PTAT core

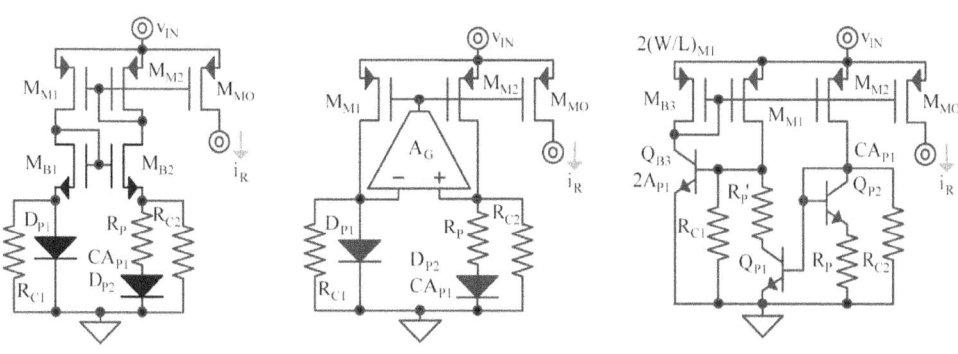

$$i_R \approx \frac{v_{D1}}{R_C} + \frac{\Delta v_{D12}}{R_P} \approx \frac{v_{D1}}{R_C} + \frac{V_t \ln C}{R_P} \quad \rightarrow \quad \text{R's should not drift much with } T_J$$

$$i_{R(BJT)} \approx \frac{v_{D1}}{R_C} + \left(\frac{V_t \ln C}{R_P}\right)\left(1 + \frac{1}{\beta_0}\right) \qquad \text{R's vary } \pm 20\% \quad \therefore \quad \text{Adjust ("trim") } R_P$$

B. i_R Example

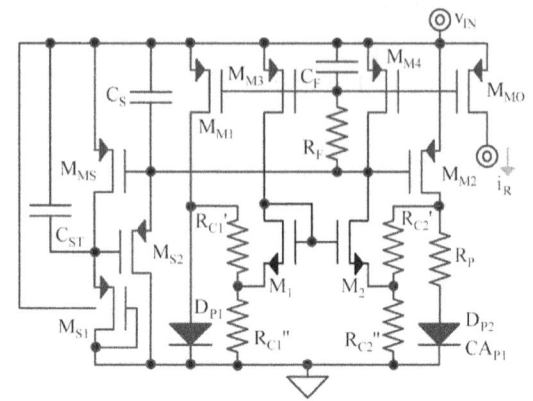

PTAT: M_M's mirror i_R

          $A_G \equiv M_{M1234}$

          R_C's voltage-divide v_D's

          R_{C2}' : A_G diode-connects M_{M2}

          R_C's : A_G matches M_{M12}'s v_D's

          D_{12} : R_P establishes v_{PTAT}

Starter: M_{S1} pulls i_{S1}

          M_{S2} pulls i_{ST} when $i_R < i_{S1}$

          C_{ST} deglitches starter

Stability: R_F : C_F reduces A_{LG+}

                C_S stabilizes – FB loop

1st-Order TC: $i_R \approx \dfrac{v_{D1}}{R_C' + R_C''} + \dfrac{V_t \ln C}{R_P}$

8.5. Voltage References: A. Inverted v_{GS}

Inverted v_{GS}: $v_{GS} = V_{T0} + V_{DS(SAT)} \approx V_{T0} + \sqrt{2\left(\dfrac{V_t \ln C}{R_P}\right)\left[\dfrac{1}{K'(W/L)_R}\right]} = v_R$

When $i_D = i_{PTAT}$ CTAT   PTAT   PTAT   CTAT   1st Order

Notes: Optimum $W/L \equiv (W/L)_R$ and R_P set the v_{PTAT} needed to balance v_{CTAT}

       → v_R = Balanced $v_{GS} \approx 1\text{–}2$ V

       C_O shunts noise

       R_P, V_{T0}, and K' can produce nonlinear drift in v_R

       →   Untrimmed Accuracy $\approx 10\%\text{–}15\%$

       i_{PTAT} biases M_R → Mono-stable → Starter is not necessary

B. Latched Core

Compact Solution: Stack PTAT core on R_{OP} $\therefore$ $v_R \approx v_{D1} + k_P i_{PTAT} R_{OP}$

Number of PTAT currents $\downarrow$

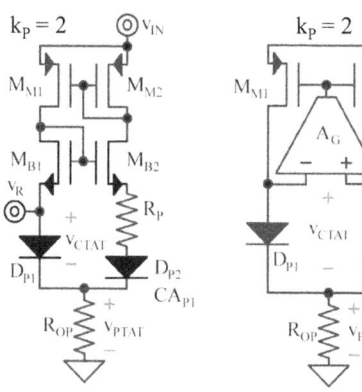
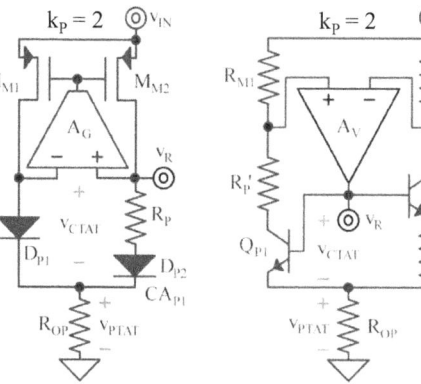
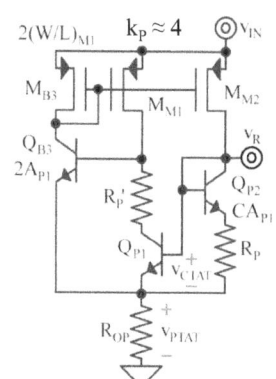

Low R_O shunts noise $\therefore$ Lower noise when v_R = Inverting shunt-FB node

$\Delta v_{D/BE}$ = Low $\rightarrow$ $v_{RP} < 100$ mV   Fractional Temp. Coefficient $\equiv \left(\dfrac{\Delta v_R}{v_R} \right)\left(\dfrac{1}{\Delta T_J} \right)$

v_{CTAT}, v_{PTAT} tolerance $\pm$ mismatches $\therefore$ Trim R_{OP} $\rightarrow$ $TC_{1st\,Order} \approx 20\text{--}100$ ppm/°C

v_R Example

1st-Order TC: $v_R \approx v_{BEP1} + 4\left(\dfrac{V_t \ln C}{R_P} \right)\left(1 + \dfrac{1}{\beta_0} \right) R_{OP}$

PTAT Core: M_M's mirror i_R

$\quad$ Trim R_{OP} Reduce T ln T with imperfections

$A_G \equiv Q_{P1B3} : M_{B3}$

A_G diode-connects M_{M2}

A_G matches M_{M12}'s v_D's & i_B's

R_P' : R_P matches v_{CE12}'s

Q_{P12} : R_P sets $i_P \approx \dfrac{v_{PTAT}}{R_P}$

Stability: R_F : C_F reduces A_{LG+}

Common-mode C_S stabilizes – FB loop

R_P boosts A_{LG-} Noise: Shunt FB reduces R_O

Starter: M_{S1} pulls i_{ST} C_O shunts noise

C. Differential CE/S Core

Concept: CE/S realization $\rightarrow$ R_P = Across bases/gates

Circuit: A_V diode-connects Q_{P1} & matches i_C's & v_{CE}'s

Q_{12} impresses $V_t \ln C$ across R_P

R_{OP} increases v_{PTAT}

A_V stacks over $R_{P/OP}$ & v_D

i_{PTAT} biases & starts circuit

Example: A_G : M_O shunt-FB v_R (regulator)

M_M's mirror i_C's R_B matches R_F's v_{RF}

Q_{P12} : R_P sets i_{PTAT} R_P voltage-divides A_{LG+}

$M_{M/O}$'s v_{SG}'s match v_{CE}'s C_S : R_S stabilizes – FB loop

R_F : C_F reduces A_{LG+} v_R Error: Q_{P2} draws i_{B2} away from R_{OP}

$$v_R \approx \left(\frac{V_t \ln C}{R_P} \right)(R_{OP} + R_P) + v_D$$

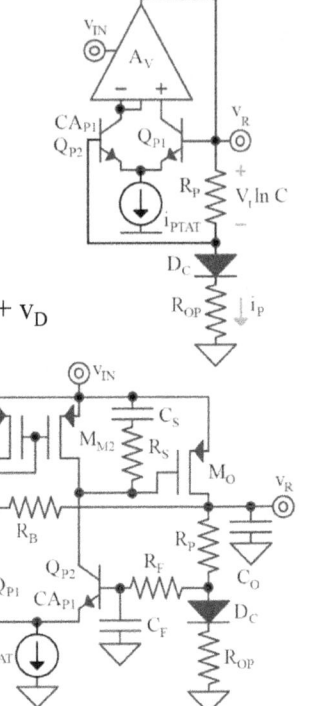

D. Sub-Bandgap

Concept: "Superimpose" sources $\rightarrow$ Current into $v_D \approx$ Voltage Source

Example: A_G : M_O shunt-FB v_R (regulator)

$$v_R \approx \frac{v_D R_O}{R_C + R_O} + i_{PTAT}(R_C \parallel R_O)$$

M_M's mirror i_C's

M_{C12} matches v_{EC}'s & i_C's

Q_{P12} : R_P sets i_{PTAT}

R_F : C_F reduces A_{LG+}

R_B matches R_F's v_{RF}

R_P voltage-divides A_{LG+}

C_S : R_S stabilizes – FB loop

$$v_R \approx V_t \ln C + \left(\frac{V_t \ln C}{R_P} \right)(R_C \parallel R_O) + \left(\frac{v_D R_O}{R_C + R_O} \right) \approx V_{BG} \left(\frac{R_O}{R_C + R_O} \right)$$

E. Tolerance

1st-Order Core: PTAT Generator → $\Delta v_{D/BE/GS}$ loop

        Current Mirror → With or without feedback amplifier

        Summer → Current or voltage mode

Error Sources: Tolerance → $v_{D/BE/GS}$ & Resistors

        Mismatch → Transistors & Resistors

        T$_J$ Drift → Resistors

Error Analysis: Deviations from ideal v_R produce Δv_R

    → $\Delta v_R = v_{R(IDEAL)} - v_{R(ERR)} = \Delta v_{CTAT} + \Delta v_{PTAT}$

Matching

Critical Pairs: PTAT Pair   Mirror Pair   PTAT Resistors   B. Cross Coupling

Goal: Average (thermal & processing) die gradients

Layout: Compact   Same Orientation (match deposition angle)

Peripheral Devices (match etching & out-diffusion effects)   (2-D)

Common Centroid ≡ Common center of mass (average 2-D gradients)

A. Inter-Digitation (1-D)

PTAT Pair: Compact   Precision

         C = 8

    C = 2   (1-D)   (2-D)

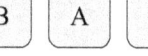

Errors

$$R_{OP} \text{ matches } R_P$$

PN Junction: $\quad i_P \approx \left(\dfrac{V_t}{R_P}\right) \ln C \qquad\qquad v_R = v_C + v_P \approx V_t \ln\left(\dfrac{i_P}{I_s}\right) + k_P V_t \left(\dfrac{R_{OP}}{R_P}\right) \ln C$

±20% Resistor Tolerance: $\qquad$ No effect on $\dfrac{R_{OP}}{R_P} \quad \therefore \quad$ No effect on v_P

Alters i_P: $\quad \Delta v_C = V_t \ln\left(\dfrac{R_P \pm \Delta R_P}{R_P}\right) = V_t \ln(1 \pm 20\%) \approx \pm 20\% \, V_t$

±1% Resistor Mismatch: $\qquad$ No effect on $i_P \quad \therefore \quad$ No effect on v_C

Alters v_P: $\quad \Delta v_P = k_P V_t (\pm 1\%) \ln C \approx \pm 4\% \, k_P V_t \quad$ (when C = 8)

Resistor T_J Drift:

Alters i_P: $\quad \Delta v_C = V_t \ln\left(\dfrac{R_{TC}}{R_P}\right) \qquad$ No effect on $\dfrac{R_{OP}}{R_P} \quad \therefore \quad$ No effect on v_P

±5% Current Mismatch: $\quad$ Produces small Δi_P

$$\text{When C = 8} \quad \left.\begin{array}{l} \\ \end{array}\right] \quad \dfrac{v_P}{V_t \ln C} \approx 12$$

Induces $\Delta v_C \approx \dfrac{\Delta i_P}{g_{mP}} = \left(\dfrac{\Delta i_P}{i_P}\right) V_t = \pm 5\% \, V_t \qquad v_P \approx 650\ \text{mV}$

Δv_D alters $i_P \quad \therefore \quad \Delta v_P = \left(\dfrac{\Delta v_C}{R_P}\right) k_P R_{OP} = \pm 5\% V_t \left(\dfrac{v_P}{\Delta v_D}\right) \approx \pm 60\% \, V_t$

Voltage Mismatch (between transistor pairs): $\quad$ Produces small Δi_P

Induces $\Delta v_C \approx \dfrac{\Delta i_P}{g_{mP}} \approx i_P \lambda_V \Delta v_M \left(\dfrac{V_t}{i_P}\right) = \lambda_V \Delta v_M V_t \qquad\qquad \lambda_V \leftrightarrow \dfrac{1}{V_A}$

Δv_D alters $i_P \quad \therefore \quad \Delta v_P = \left(\dfrac{\Delta v_C}{R_P}\right) k_P R_{OP} = \lambda_V \Delta v_M V_t \left(\dfrac{v_P}{\Delta v_D}\right) \qquad\qquad 13\, \lambda_V \Delta v_M V_t$

CMOS Sub-v_T: $\quad V_t \to \eta_I V_t \quad \therefore \quad \Delta v_R$ is η_I times (up to 2×) higher

Package Stress (shift)

Piezoelectric Effect ≡ Electrical effects of mechanical stress

Plastic expands with T_J & melts at 170°C → Contracts when T_J falls

∴ Systemic stress on die that relaxes with rising T_J → $\Delta v_{R(S)}$ fades with T_J

TCoE ≡ Thermal coefficient of expansion

Package's TCoE >> Silicon die's TCoE → Die can crack when T_J falls

∴ Add plastic fillers with lower TCoE (to reduce package's overall TCoE)

   Fillers exert random local stresses → Δv_R^*

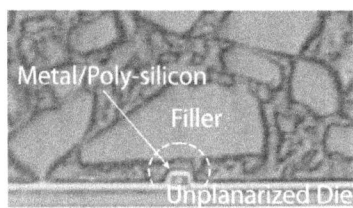
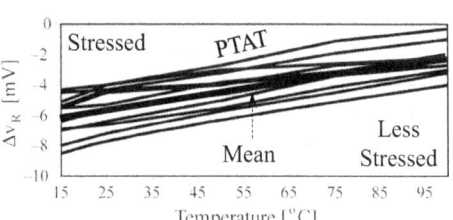

Stress Reduction

Systemic Component

$\Delta v_{R(S)} \propto$ PTAT → Model in simulations & compensate with lower v_{PTAT}

Random Component   Extra Layers/Ceramic Package: Higher cost

"Cushion" fillers with mechanically compliant layers:

Uniform metal planes   Nitride layer   Post-fabrication layers

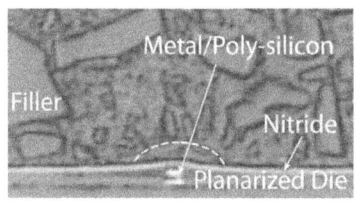
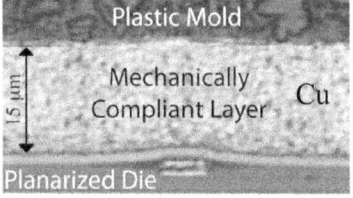

16%–18% lower Δv_R^* 35% lower Δv_R^*

Ceramic Package: Low TCoE → Low piezoelectric effects (package shift)

Baseline Tolerance

Error Source		Error	Drift	Δv_R
20% R Tolerance		$\pm 20\%\ V_t$	PTAT	± 5 mV
1% R Mismatch ($v_P = 2 i_P R_{OP}$)		$\pm 4\%\ k_P\ V_t$		± 2 mV
R Drift		$V_t \ln \%\Delta R$		Can be low
Current Mismatch	PTAT Pair	$\pm 5\%\ (12)\ V_t$	PTAT	± 15 mV
	Mirror	$\pm 5\%\ (12)\ V_t$		± 15 mV
Voltage Mismatch		$13\ \lambda_V\ \Delta v_M\ V_t$		Can be low
v_D Tolerance		$\pm 2\%$		± 24 mV
Total Δv_R				0 ± 32 mV
Package Stress		Unplanarized	PTAT	-6 ± 12 mV
Total Δv_R				-6 ± 35 mV

Tolerance: $\Delta v_R = \pm 32$ mV of $v_R \approx V_{BG}$ is $\pm 2.7\%$ accuracy

Trim: Adjusting v_{PTAT} with R_{OP} can compensate PTAT errors

www.ingramcontent.com/pod-product-compliance
Lightning Source LLC
Chambersburg PA
CBHW081430220526
45466CB00008B/2331